RESEARCH AND WRITING
ACROSS THE DISCIPLINES

RESEARCH AND WRITING
ACROSS THE DISCIPLINES

P Ramadass

Professor and Head (Retd.)
Department of Animal Biotechnology
Madras Veterinary College
Tamil Nadu Veterinary and Animal Sciences University
Chennai

A Wilson Aruni

Associate Professor
Department of Veterinary Microbiology
Tamil Nadu Veterinary and Animal Sciences University
Chennai

MJP PUBLISHERS

MJP Publishers

New No. 5 Muthu Kalathy Street,
Triplicane,
Chennai 600 005

MJP 059   © Publishers,

Publisher : J.C. Pillai

PREFACE

This book was conceived as a result of many years of our experience in teaching research methodology for postgraduate students. The purpose of this book is to provide the research students, mainly postgraduate students, with a summary of important steps to be used for selection of research topics, selection of research methods, analysis of results and arriving at conclusions and writing up the thesis or project reports. The book is designed to help students learn to develop capabilities for formulation of thesis/research topics, execution of the research and effective writing of the research results. At the college level, the effective thinker, researcher, and writer possesses at least the following cognitive skills:

- the ability to analyse the available literature and to find out a suitable research topic
- the ability to identify the appropriate research methodology for carrying out the experiments
- the ability to analyse the results and bring out conclusions
- the ability to write the thesis/dissertation/research report
- the ability to write a research proposal or research grant proposal

This book encourages students to develop these abilities not as ends in themselves, but as means to becoming self-generating thinkers, researchers and writers. It is not about simple thesis topic selection and conduct of research, but, the whole exercise of going through the vast amount of literature available in a particular topic, finding out the most appropriate research method, analysing the results, arriving at conclusions and putting them in the forms of a presentable thesis format.

In Part I of this book, students are presented with different research methods, research designs, data collection and analysis methods. Part II chapters deal with use of library for literature collection, selection of research topic, writing different sections of thesis, proofreading, use of photography for photo-documentation of results, writing of research proposals and research grant proposals. Finally, organization of meetings, discussions and seminars are given.

We intend that the book should be used either as a textbook on Research Methodology or as an easy reference for research students who are planning to undertake a project work/dissertation/thesis work for award of a degree. We hope that the book will be user-friendly for research students who are venturing into research for the first time.

We would welcome feedback from readers regarding any errors of either omission or fact that may inadvertently be present, and would welcome any suggestions as to how the book's usefulness may be improved in the future.

P. Ramadass

A. Wilson Aruni

CONTENTS

PART I

ESSENTIALS OF RESEARCH

1

AN INTRODUCTION TO RESEARCH METHODOLOGY

RESEARCH METHODS VS RESEARCH METHODOLOGY

Research methods may be understood as all those methods/ techniques that are used for conducting research. Research methods or techniques refer to the methods the researchers use in performing research operations. In other words, all those methods which are used by the researcher during the course of studying their research problem are termed research methods. Research methods can be put into the following three groups.

1. In the first group, we include those methods which are concerned with the collection of data.

2. The second group consists of those statistical techniques which are used for establishing relationships between the data and the unknowns.

3. The third group consists of those methods which are used to evaluate the accuracy of the results obtained.

Research methodology is a way to systematically solve the research problem. It may be understood as a science of studying how research is done scientifically. In it, we study the various steps that are generally adopted by a researcher in studying his research problem along with the logic behind them. It is necessary for the researcher to know not only the research methods/techniques, but also which of these methods are relevant and which are not, and what they should mean and indicate and why. Researchers also need to understand the assumptions underlying various techniques and to know the criteria by which they can decide that certain techniques and procedures will be applicable to certain problems and others will not. All this means that it is necessary for the researcher to design the methodology for their problem as the same may differ from problem to problem.

Research methodology has many dimensions and research methods do constitute a part of the research methodology. The scope of research methodology is wider than that of research methods. Thus, when we talk of research methodology, we not only talk of research methods, but also consider the logic behind the methods we use in the context of our research study, and explain why we are not using others so that research results are capable of being evaluated either by us or by others. Why a research study has been undertaken, how the research problem has been defined, in what way and why the hypothesis has been formulated, what data have been collected and what particular method has been adopted, why a particular technique of analysing data has been used and a host of similar other questions are usually answered when we talk of research methodology concerning a research problem or study.

WHAT IS RESEARCH?

Research is a scientific search for pertinent information on a specific topic. Research is an art of scientific investigation. Fact finding is research. Research refers to a search for knowledge. Scientific and systemic search for pertinent information on a specific topic is research. Research is a piece of work done to find out an answer to a question either hypothetical or real. Research is a systematic investigation, involving the collection of information (data), to solve a problem or contribute to knowledge about a theory or practice. It applies the scientific approach to the study of a question of interest and relies on methods and principles that will produce credible and verifiable results. Research helps provide scientific understanding and solves practical problems. Research is the honest desire to know something coupled with an energetic search to find the answer. It is an honest desire to explore, observe or analyse a subject. Its reward should be the satisfaction that comes with the knowing.

Scientific research involves a careful planning, an orderly execution and an effective way of presentation of results and drawing conclusions with reference to past work. Research in any one of the scientific fields means an intensive and purposeful search for knowledge and understanding of scientific phenomena. It is not only a method for the discovery of true values in scientific knowledge, but also a critical and scientific analysis of facts and formulations of generalization as a basis for action and foresight. The general criteria of the scientific method consist of systematic observations, verification and classification of the facts, and the interpretation and generalization about possible relations.

STEPS IN RESEARCH (SCIENTIFIC APPROACH)

◎ Identify the question or problem

◎ Conduct a literature review of the question

◎ Frame a hypothesis

◎ Develop the study design

◎ Complete the research

◎ Analyse the data

◎ Generate conclusions

The purpose of the research is to discover new facts and in the light of newly discovered facts, enable the scientist to confirm, revise or reject the existing theories and accepted practices. The research is an act involving appreciation and expression. It helps the researcher to acquire a thorough knowledge of the topic under investigation. Not only does the researcher get practice in appraising materials, in giving correct interpretation and in drawing conclusions, but also develops his/her skills in writing and effective expression.

"Knowledge is of two kinds: we know a subject ourselves; or we know where we can find information upon it." (Boswell's Life of Johnson, 1775).

NEED FOR RESEARCH IN DEVELOPING COUNTRIES

Research involves a drain on resources. Hence, in developing countries, where resources crunch is omnipresent, planners always question the wisdom of drawing the meagre resources for research, and argue that the results of corresponding research from developed countries could be adopted. As long as the conditions are exactly identical, there is no harm in adopting the results of research conducted elsewhere. But given the diversity in nature and otherwise, we always find variations in the conditions under which a research work was carried out elsewhere. We find variations in the environment such as agro-climatic conditions, geographical conditions and diversity in the fauna and flora, besides the greatest diversity in the social set-up and the consequent practice. Hence, more often than not, it becomes necessary to repeat the research carried out in developed countries under conditions prevalent in the developing countries. Therefore, there is the need for research in developing countries, which depends on a variety of factors.

ROLE OF RESEARCH IN SCIENTIFIC ADVANCEMENT

Research is essentially done to further our knowledge in our field of interest. For example, to know about the disease causing agents and the pathogenesis of diseases and for controlling the diseases by newly developed high-tech vaccines, understanding the biology and molecular structures of the microbes will help to develop methods to prepare newer vaccines to control the diseases. Knowing different structural proteins of viruses like the foot-and-mouth disease virus (like VP1 protein, especially amino acid residues 141–161 and 200–213) has helped to develop recombinant subunit vaccines, which are safer than the conventional vaccines. In rotavirus, VP7 fraction has been used for vaccine production.

Research in the pathogenesis of a disease will help to understand the disease process in a host, which will eventually be used to diagnose the disease in the earlier stage of infection. Research in the molecular structures of microbes will lead to the development of nucleic acid probes, which could be used in diagnosis of the disease. Research into the chromosome and genes have helped to identify the genetic defects at the earlier period of development of an animal. For example, with the aid of karyotyping studies in bovines, Robertsonian translocation has been identified, which induces offspring which are mostly sterile and unfit for breeding.

Research at the genetic level to identify point mutation, which may result in genetic disorders like endemic goitre, sickle cell anaemia could be detected at earlier stage of development. RFLP analysis has helped to identify various genetic defects. Research in the physiology of rumen microbes may help to improve the efficiency of rumen microbes for efficient feed conversion. Research into the effects of drugs on various organs of the body will lead to the development of newer drugs with more efficiency and lesser toxic effects to the body. Research in the methods of artificial induction of superovulation has lead to the development of embryo transfer technology.

OBJECTIVES OF RESEARCH

Clear-cut objectives are to be set out before taking any research work. While formulating a research project, the need for research work has to be analysed. How the outcome of the research work will benefit the society is to be analysed. The following questions are to be asked before taking any research work:

1. Why a research work has to be done?
2. Where the research work should be undertaken?
3. When the research work should be undertaken?
4. How a research work should be done?
5. By whom the research work should be done?

Why?—Reasons for research work. Taking a particular problem in a society, e.g. epidemiological studies on the occurrence of gastroenteritis.

Where?—Select a population where there is a problem. An outbreak of gastroenteritis in a locality. Go in a methodical way to collect information and samples for analysis mainly from low-income groups. For research on effect of alcohol on health, information is to be collected from the high-income group.

When?—Finding out the right time of research is important. Gastrointestinal disorders are more common during the rainy season. For studying problems associated with calving, work has to be done during the calving season. Mosquito-borne infections are studied after the rainy season.

How?—For carrying out research work, proper methodology has to be adopted. The number of samples to be analysed should be determined. Statistical analysis has to be performed and the period of investigation, etc. has to be planned.

By whom?—Research work should be carried out by competent persons. Otherwise, results obtained cannot be useful to any one. Persons who know some basic things about the subject should be entrusted with the research work. One should be sincere and dedicated to do research work and record the results meticulously and methodically.

There are four main types of research:

1. To gain familiarity with a phenomenon or to achieve new insight into it (**exploratory or formulative research studies**).

2. To portray accurately the characteristics of a particular individual, situation or a group (**descriptive research studies**).

3. To determine the frequency with which something occurs or with which it is associated with something else (**diagnostic research studies**).

4. To test a hypothesis of a causal relationship between variables (**hypothesis testing research studies**).

MOTIVATION IN RESEARCH

What makes people to undertake research? The possible motives for doing research may be either one or more of the following:

i. Desire to get a research degree along with its consequential benefits;

ii. Desire to face the challenge in solving the unresolved problems;

iii. Desire to get intellectual joy of doing some creative work;

iv. Desire to be of service to society;

v. Desire to be respected and

vi. Directives of the government, employment conditions, etc.

TYPES OF RESEARCH

Descriptive vs Analytical Research

Descriptive research includes survey and fact-finding enquiries of different kinds. The major purpose of descriptive research is description of the state of affairs as it exists at present. The main feature of this method is that the researcher has no control over the variables; they can report what has happened or what is happening. Survey methods of all kinds including comparative and correlational methods are the methods of research utilized in descriptive research. In **analytical research**, the researcher has to use the facts or information already available and analyse these to make a critical evaluation of the materials.

Applied vs Fundamental Research

Research can either be **applied** (or action) or **fundamental (or basic) research**. Applied research aims at finding a solution for an immediate problem facing society or industry, whereas the fundamental research is mainly concerned with generalization and with the formulation of a theory. "Gathering knowledge for knowledge's sake is termed as basic research". Research concerning taxonomical classification of bacterial species by DNA homology studies is an example of fundamental research, whereas, research on the efficiency of vaccines or drugs is an example of applied research.

Qualitative vs Quantitative Research

Quantitative research is based on the measurement of quantity or amount. Milk yield in different breeds of cows over a period of time refers to quantitative research. It is applicable to phenomena that can be expressed in terms of quantity. **Qualitative research**, on the other hand, is concerned with qualitative phenomena, i.e., phenomena relating to or involving quality or kind, e.g. qualities of milk like flavour, colour, etc. When we are interested in investigating a reason for human behaviour (why people think or do certain things), we quite often talk of motivation research, an important type of qualitative research.

Conceptual vs Empirical Research

Conceptual research is that related to some abstract idea(s) or theory. It is generally used by philosophers or thinkers to develop a new concept or to reinterpret existing ones. On the other hand, **empirical research** relies on experience or observation alone, often without due regard for the system or theory. It is data-based research, coming up with conclusions which are capable of being verified by observation or experiment. We can also call it experimental type of research. Empirical research is appropriate when proof is sought with certain variables that affect other variables in some way. Evidence gained through experiments or empirical studies is considered to be the most powerful support for a given hypothesis.

Other Types of Research

From the point of view of time, research can be a **one-time research** or **longitudinal research**. In the former case, a research is confined to a single time period, whereas in the latter case, the research is carried on over several time periods. Research can be **field-setting research** or **laboratory research** or **simulation research**, depending upon the environment in which it is to be carried out. Research can also be **clinical** or **diagnostic research**. Such research follows case study methods or in-depth approaches to reach the basic causal relations.

Research may also be **exploratory** or may be **formalized**. The objective of exploratory research is the development of a hypothesis rather than testing, whereas formalized research studies are those with substantial structure and with a specific hypothesis to be tested. Research can be **historical research**, which utilizes historical sources like documents, remains, etc. to study events or ideas of the past. A research may also be a **conclusion-oriented research**, where a researcher is free to pick up a problem, redesign the enquiry as they proceed and be prepared to conceptualize as they wish. But a **discussion-oriented research** is always for the need of the decision-makers, and a researcher in this is not free to embark upon research according to theirs own inclination.

TECHNIQUES OF RESEARCH

The determination of the research technique is closely associated with the purpose of the research project and resources available. One of the important uses of research techniques is to provide a means to infer characteristics of a large group of objects from the study of a subgroup of these objects. The result of such a study is called **observation**. The subgroup that is actually studied is called **sample**. Research is almost always directed at characterizing and understanding a segment of a population. A representative sample tends to contain relevant characteristics in the sample population as they exist in the population.

Complete enumeration and sampling are two techniques in which the researcher may obtain the required information. Under complete enumeration survey, each and every unit of the population is covered and contacted. Sampling is the process of learning about the population on the basis of samples drawn from it. Complete enumeration technique is very rarely used. Sampling techniques are becoming more popular in research investigations. Sampling techniques involve reduced cost, higher speed, wider scope and greater accuracy.

Types of Sampling

Sampling techniques can be grouped under two broad heads: 1) random sampling and 2) non-random sampling.

Random sampling   Also known as **chance sampling** or **probability sampling**, is the form of sampling applied when the method of selection assures each individual or element an equal chance of being selected. Random sampling is an essential part of protection against bias and prejudice. Random sampling has also been called proportionate sampling because each class of item is in the same proportion in the sample as in the universe. In order to evolve a satisfactory technique of pure random sampling, personal choice must be completely eliminated and proper procedure based on the theory of probability should be adopted. Some of the commonly followed techniques for selection of random items are as follows:

i. **Lottery method**   Numbers or names are written on chits, capsules or balls. They are put in a container and thoroughly mixed. The required numbers of chits are drawn from the container.

ii. **Random sampling numbers**   Tables of numbers which have been thoroughly tested for randomness are now available for selection of such samples. A table of true random numbers is one in which digits from 0 to 9 have an equal chance of appearing in any position of the table. A random series constructed by L.B.C. Tippet consists of 10,400 four digit numbers.

1. *Stratified random sampling* If population does not constitute a homogeneous group, this method is adopted. The field of enquiry in investigation may be composed of items heterogeneous in character. Under such circumstances, reliable conclusions cannot be inferred by selecting samples either on a random basis or through purposive randomization. The only alternation in a heterogeneous population is to select samples after proper stratification of the universe. The population is stratified into a number of subpopulations or strata and sample is selected from each stratum. If the items selected from each stratum are based on simple random sampling, it is known as stratified random sampling, e.g. Fat percentages of milk of cattle and buffaloes and incidence of gastro-intestinal infection in children and adults.

2. *Purposive or deliberate sampling* Also known as **non-probability sampling**, **judgement sampling** or **convenience sampling**, this is not considered representative because, the method of selecting samples for investigation is subjective. Here, the researcher selects the units of samples that are not representative of the population, i.e., they are subjective. The selection of samples can hardly be random and unbiased, and the personality of the selector enters into selection. This is also known as judgement sampling because judgement is used for selecting items which the researcher considers as representative of the population.

3. *Systematic sampling* Selecting every 15th sample on a list, every 10th house on one side of a street and so on could be examples of systematic sampling. An element of randomness is introduced to pick up the unit with which to start.

4. *Cluster sampling and area sampling* This involves grouping the population and then selecting the group or the clusters rather than individuals. In area sampling, the total area is divided into smaller areas and a number of these smaller areas are randomly selected.

5. *Multistage sampling* This is meant for big enquiries to a considerably larger geographical group like an entire country. In this method, in the first stage, large primary sampling units such as states are selected, then districts, towns and finally certain families within towns may be selected. If the technique of random sampling is applied at all stages, the sampling procedure is described as multistage random sampling.

Non-random sampling   Non-random (or "non-probability") samples are selected by any kind of procedure that does not give all cases in the population equal chances to fall into the sample. Sometimes the context of the study allows or facilitates using a certain method of sampling; sometimes the researcher has the possibility of selecting the method. This method of sampling will favour certain types of cases in the population more than the others, in other words the data from the sample will be biased. When assessing a non-random sample you should ask yourself: Will the results from the sample be the same that you would get from the population? Is it certain that the criterion that you have used in selecting the sample (e.g. the willingness of people to participate) has no relationship with those variables that you want to record from the sample? If there is correlation, your sample will be biased and you should consider constructing a new sample with less correlation.

Common types of non-random sampling include:

Convenience sampling   It simply involves using the people who are the most available or the most easily selected to be in your research study. A coincidental group, e.g. people at a meeting, might be specified as a sample. More exactly, the sample contains those persons in the group who are willing to take part. Such a sample is often heavily biased, but this can be accepted if the data obtained from the sample are not really going to be used, such as is the case in demonstrations in survey method classes at universities.

Quota sampling   Another type of non-random sampling is called quota sampling (i.e., it involves setting quotas and then using convenience sampling to obtain those quotas). A set of quotas might be given to you as follows: find 25 African American males, 25 European American males, 25 African American females, and 25 European American females. You use convenience sampling to actually find the people, but you must make sure you have the right number of people for each quota.

Purposive sampling   The researcher specifies the characteristics of the population of interest and then locates individuals who match those characteristics. For example, you might decide that you want to only include "boys who are in the 7th grade and have been diagnosed with malnutrition" in your research study. You would then, try to find 50 students who meet your "inclusion criteria" and include them in your research study.

Snowball sampling   Each research participant is asked to identify other potential research participants who have a certain characteristics. You start with one or a few participants, ask them for more, find those, ask them for some, and continue until you have a sufficient sample size. This technique might be used for a hard-to-find population (e.g. where no sampling frame exists). For example, you might want to use snowball sampling if you wanted to do a study of people in your city who have a lot of power in the area of educational policy making (in addition to the already known positions of power, such as the school board and the school system superintendent). Snowball sampling is a good method for such populations that are not well delimited nor well enumerated, for example the homeless. The drawback is that you get no exact idea of the factual distribution of the opinions in the target population. Besides, people usually propose people that they know well and who share their own views, which means that small groups of interest often are passed by unnoticed.

Sample of volunteers   This is created when all the members of a population have the opportunity to participate in the sample, and all volunteers are accepted. If you insert a survey form in an Internet page and ask people to give their opinions on a topic, you will get this type of sample from all the readers of this page. Similarly, the persons who spontaneously send customer feedback to company are a sample of volunteers from all the customers. A sample of volunteers can be a practical alternative when there is no list of the members of the population from which a random sample could be drawn, or when it is difficult to contact the people in a sample because their addresses are not known. The disadvantage is that it is difficult to assess the presence of bias, i.e., whether the opinions or other interesting properties of the volunteers deviate from those of the population.

Sample that consists of all the available cases   Sometimes the researcher is interested in a population of which only a few cases or specimens are available for study, and these then must serve as a sample of the population. Typical such samples are:

- Surviving cases
- Permitted cases

Surviving cases   Among historical or archaeological material, when a large part of once relevant material have disappeared before researchers get at it, can be regarded as a kind of convenience sample even when it is the historical reality and not convenience that selects the sample. Both samples involve a similar handicap for research: if the disappearance of material, during the period until the study, has not been random nor proportional but instead somehow partial or selective, the remaining material will be biased and the researcher should try to assess the likely bias.

Permitted cases   When studying private enterprises it often happens that the management will not allow recording information from certain units in the organization. The management's decision perhaps is motivated by their judgment about the objectives of the study, but from the researcher's scientific point of view such a sample will often seem seriously biased.

Sizes of Samples

The size of the samples has an indirect relation with the degree of accuracy desired in research. It is essential that sample size should be carefully chosen. It has been the general experience that larger the size of samples, the greater is the degree of accuracy.

Adequacies of Samples

Adequacy of samples is based on representativeness and precision. The degree of accuracy and precision depends on the size of the sample and an increase in the size of the sample number reduces the effect of chance errors of sampling. A sample result based on a few abnormalities of measurements may not be adequate because it may lack the precision necessary for an investigation. Similarly, biased samples are never adequate because they are hardly representative of the sample.

QUALIFICATIONS OF A RESEARCHER

1. **Passion for original and creative work**   A research worker should be seeking ever newer truths in the interpretation of scientific phenomena. Research interest develops only when there is a craving for discovery and zeal for investigation into fields of human inquiry.

2. **Steady and incision observation**   A researcher's mind should be versatile enough for observing subtle shades of difference,

for getting straight to the point and for understanding the causal relations.

3. **Capacity for solving problems** A researcher should have the capacity for solving problems, both deductive and inductive. They should develop their reasoning power to attain accuracy. They should process a sound sense of logic so that they are perspicacious in making deductions.

4. **Continuous concentration in reading and analysis** A research worker should have sustained interest in exploring new fields of knowledge. They should be sincere, single-minded and much earnest in pursuing their efforts until success is achieved.

5. **Accuracy** This is the hallmark of efficiency in research. A research worker should make use of statistical methods for presenting their data.

6. **Intellectual honesty and wholesome moral standards** They should be honest and reliable not only in the collection, but also in the presentation of facts.

7. **Clear and effective expression** The composition of the research report should be brief, exact and clear, so that the reader may be able to grasp the main idea and the general effect of the matter presented.

8. **Budgeting the time** A painstaking planned work is an important requisite for any research activity. Just as the selection of a proper subject or problem for the research study, determination of time for completing the study is also no less an important factor. The researchers should be able to allot some rough time for the various phases of their work.

2

THE SCIENTIFIC METHOD

INTRODUCTION

The scientific method is the process by which scientists, collectively and over time, endeavour to construct an accurate representation (that is, reliable, consistent and non-arbitrary) of the world.

Recognizing that personal and cultural beliefs influence both our perceptions and our interpretations of natural phenomena, we aim through the use of standard procedures and criteria to minimize those influences when developing a theory. As a famous scientist once said, "Smart people (like smart lawyers) can come up with very good explanations for mistaken points of view." In summary, the scientific method attempts to minimize the influence of bias or prejudice in the experimenter when testing a hypothesis or a theory.

STEPS OF THE SCIENTIFIC METHOD

The scientific method has four steps:

1. Observation and description of a phenomenon or group of phenomena.
2. Formulation of a hypothesis to explain the phenomena. In physics, the hypothesis often takes the form of a causal mechanism or a mathematical relation.
3. Use of the hypothesis to predict the existence of other phenomena, or to predict quantitatively the results of new observations.
4. Performance of experimental tests of the predictions by several independent experimenters and properly performed experiments.

If the experiments bear out the hypothesis it may come to be regarded as a theory or law of nature. If the experiments do not bear out the hypothesis, it must be rejected or modified. What is key in the description of the scientific method just given is the predictive power of the hypothesis or theory, as tested by experiment. It is often said in science that theories can never be proved, only disproved. There is always the possibility that a new observation or a new experiment will conflict with a long-standing theory.

TESTING HYPOTHESES

As just stated, experimental tests may lead either to the confirmation of the hypothesis, or to the ruling out of the hypothesis. The scientific method requires that a hypothesis be ruled out or modified if its predictions are clearly and repeatedly incompatible with experimental tests. Further, no matter how elegant a theory is, its predictions must agree with experimental results if we are to believe that it is a valid description of nature. In physics, as in every experimental science, "experiment is supreme" and experimental verification of hypothetical predictions is absolutely necessary. Experiments may test the theory directly (for example, the observation of a new particle) or may test for consequences derived from the theory using mathematics and logic (the rate of a radioactive decay process requiring the existence of the new particle). Note that the necessity of experiment also implies that a theory must be testable. Theories which cannot be tested, because, for instance, they have no observable ramifications (such as, a particle whose characteristics make it unobservable), do not qualify as scientific theories.

If the predictions of a long-standing theory are found to be in disagreement with new experimental results, the theory may be discarded as a description of reality, but it may continue to be applicable within a limited range of measurable parameters. For example, the laws of classical mechanics (Newton's Laws) are valid only when the velocities of interest are much smaller than the speed of light (that is, in algebraic form, when $v/c << 1$). Since this is the domain of a large portion of human experience, the laws of classical mechanics are widely, usefully and correctly applied in a large range of technological and scientific problems. Yet in nature we observe a domain in which v/c is not small. The motions of objects in this domain, as well as motion in the "classical" domain, are accurately described through the equations of Einstein's theory of relativity. We believe, due to experimental tests, that relativistic theory provides a more general, and therefore more accurate, description of the principles governing our universe, than the earlier "classical" theory. Further, we find that the relativistic equations reduce to the classical equations in the limit $v/c << 1$. Similarly, classical physics is valid only at distances much larger than atomic scales ($x >> 10^{-8}$ m). A description which is valid at all length scales is given by the equations of quantum mechanics.

We are all familiar with theories which had to be discarded in the face of experimental evidence. In the field of astronomy, the earth-centred description of the planetary orbits was overthrown by the Copernican system, in which the sun was placed at the centre of a series of concentric, circular planetary orbits. Later, this theory was modified, as measurements of the planets motions were found to be compatible with elliptical, not circular, orbits, and still later planetary motion was found to be derivable from Newton's laws.

Error in experiments have several sources. First, there is error intrinsic to instruments of measurement. Because this type of error has equal probability of producing a measurement higher or lower numerically than the "true" value, it is called random error. Second, there is non-random or systematic error, due to factors which bias the result in one direction. No measurement, and therefore no experiment, can be perfectly precise. At the same time, in science we have standard ways of estimating and in some cases reducing errors. Thus it is important to determine the accuracy of a particular measurement and, when stating quantitative results, to quote the measurement error. A measurement without a quoted error is meaningless. The comparison between experiment and theory is made within the context of experimental errors. Scientists ask, how many standard deviations are the results from the theoretical prediction? Have all sources of systematic and random errors been properly estimated?

COMMON MISTAKES IN APPLYING THE SCIENTIFIC METHOD

As stated earlier, the scientific method attempts to minimize the influence of the scientist's bias on the outcome of an experiment. That is, when testing a hypothesis or a theory, the scientist may have a preference for one outcome or another, and it is important that this preference does not bias the results or their interpretation. The most fundamental error is to mistake the hypothesis for an explanation of a phenomenon, without performing experimental tests. Sometimes "common sense" and "logic" tempt us into believing that no test is needed. There are numerous examples of this, dating from the Greek philosophers to the present day.

Another common mistake is to ignore or rule out data which do not support the hypothesis. Ideally, the experimenter is open to the possibility that the hypothesis is correct or incorrect. Sometimes, however, a scientist may have a strong belief that the hypothesis is true (or false), or feels internal or external pressure to get a specific result. In that case, there may be a psychological tendency to find "something wrong", such as systematic effects, with data which do not support the scientist's expectations, while data which do agree with those expectations may not be checked as carefully. The lesson is that all data must be handled in the same way.

Another common mistake arises from the failure to estimate quantitatively systematic errors (and all errors). There are many examples of discoveries which were missed by experimenters whose data contained a new phenomenon, but who explained it away as a systematic background. Conversely, there are many examples of alleged "new discoveries" which later proved to be due to systematic errors not accounted for by the "discoverers."

In a field where there is active experimentation and open communication among members of the scientific community, the biases of individuals or groups may cancel out, because experimental tests are repeated by different scientists who may have different biases. In addition, different types of experimental set-up have different sources of systematic errors. Over a period spanning a variety of experimental tests (usually at least several years), a consensus develops in the community as to which experimental results have stood the test of time.

HYPOTHESES, MODELS, THEORIES AND LAWS

In physics and other science disciplines, the words "hypothesis," "model," "theory" and "law" have different connotations in relation to the stage of acceptance or knowledge about a group of phenomena.

A hypothesis is a limited statement regarding cause and effect in specific situations; it also refers to our state of knowledge before experimental work has been performed and perhaps even before new phenomena have been predicted. To take an example from daily life, suppose you discover that your car will not start. You may say, "My car does not start because the battery is low." This is your first hypothesis. You may then check whether the lights were left on, or if the engine makes a particular sound when you turn the ignition key. You might actually check the voltage across the terminals of the battery. If you discover that the battery is not low, you might attempt another hypothesis ("The starter is broken"; "This is really not my car.")

The word model is reserved for situations when it is known that the hypothesis has at least limited validity. An often cited example of this is the Bohr model of the atom, in which, in an analogy to the solar system, the electrons are described as moving in circular orbits around the nucleus. This is not an accurate depiction of what an atom "looks like," but the model succeeds in mathematically representing the energies (but not the correct angular momentum) of the quantum states of the electron in the simplest case, the hydrogen atom. Another example is Hook's Law (which should be called Hook's principle, or Hook's model), which states that the force exerted by a mass attached to a spring is proportional to the amount the spring is stretched. We know that this principle is only valid for small amounts of stretching. The "law" fails when the spring is stretched beyond its elastic limit (it can break). This principle, however, leads to the prediction of simple harmonic motion, and, as a model of the behaviour of a spring, has been versatile in an extremely broad range of applications.

A scientific theory or law represents a hypothesis, or a group of related hypotheses, which has been confirmed through repeated experimental tests. Theories in physics are often formulated in terms of a few concepts and equations, which are identified with "laws of nature," suggesting their universal applicability. Accepted scientific theories and laws become part of our understanding of the universe and the basis for exploring less well-understood areas of knowledge. Theories are not easily discarded; new discoveries are first assumed to fit into the existing theoretical framework. It is only when, after repeated experimental tests, the new phenomenon cannot be accommodated that scientists seriously question the theory and attempt to modify it.

The validity that we attach to scientific theories as representing realities of the physical world is to be contrasted with the facile invalidation implied by the expression, "It's only a theory." For example, it is unlikely that a person will step off a tall building on the assumption that they will not fall, because "Gravity is only a theory."

Changes in scientific thought and theories occur, of course, sometimes revolutionizing our view of the world. Again, the key force for change is the scientific method, and its emphasis on experiment.

CIRCUMSTANCES IN WHICH THE SCIENTIFIC METHOD IS NOT APPLICABLE

While the scientific method is necessary in developing scientific knowledge, it is also useful in everyday problem-solving. What do you do when your telephone doesn't work? Is there a problem in the hand set, in the cabling inside your house, in the hook-up outside, or in the workings of the phone company? The process you might go through to solve this problem could involve scientific thinking, and the results might contradict your initial expectations.

Like any good scientist, you may question the range of situations (outside of science) in which the scientific method may be applied. From what has been stated above, we determine that the scientific method works best in situations where one can isolate the phenomenon of interest, by eliminating or accounting for extraneous factors, and where one can repeatedly test the system under study after making limited, controlled changes in it.

There are, of course, circumstances when one cannot isolate the phenomena or when one cannot repeat the measurement over and over again. In such cases the results may depend in part on the history of a situation. This often occurs in social interactions between people. For example, when a lawyer makes arguments in front of a jury in court, she or he cannot try other approaches by repeating the trial over and over again in front of the same jury. In a new trial, the jury composition will be different. Even the same jury hearing a new set of arguments cannot be expected to forget what they heard before.

CONCLUSION

The scientific method is intricately associated with science, the process of human inquiry that pervades the modern era on many levels. While the method appears simple and logical in description, there is perhaps no more complex question than that of knowing how we come to know things. In this introduction, we have emphasized that the scientific method distinguishes science from other forms of explanation because of its requirement of systematic experimentation. We have also tried to point out some of the criteria and practices developed by scientists to reduce the influence of individual or social bias on scientific findings.

3

TYPES OF RESEARCH STUDIES

The qualitative versus quantitative approach to the classification of research activities classifies all research studies into one of six categories.

Qualitative approach involves the collection of extensive narrative data in order to gain insight into the phenomena of interest. Data analysis includes the coding of the data and production of a verbal synthesis (inductive process).

1. Historical research
2. Qualitative research

Quantitative approach involves the collection of numerical data in order to explain, predict, and/or control phenomena of interest. Data analysis is mainly statistical (deductive process).

1. Descriptive research
2. Correlational research
3. Causal-comparative research
4. Experimental research

QUALITATIVE RESEARCH APPROACHES

Historical research and qualitative research are the two types of research classified as qualitative research approaches.

Historical research is involved with the study of past events.

The following are some examples of historical research studies mentioned by Gay.

   i. Factors leading to the development and growth of cooperative learning.

   ii. Effects of decisions of the United States Supreme Court on American education.

   iii. Trends in reading instruction, 1940–1945.

Qualitative research, also referred to as ethnographic research, is involved in the study of current events rather than past events. It involves the collection of extensive narrative data (non-numerical data) on many variables over an extended period of time in a naturalistic setting. Participant observation, where the researcher lives with the subjects being observed is frequently used in qualitative research. Case studies are also used in qualitative research.

Some examples of qualitative studies mentioned by Gay are:

   i. A case study of parental involvement at a management school.

   ii. A multicase study of students who excel despite nonfacilitating environments.

   iii. The teacher as researcher: Improving students' writing skills.

Qualitative research approaches began to gain recognition in the 1970s. The phrase "qualitative research" was until then marginalized as a discipline of anthropology or sociology, and terms like ethnography, fieldwork, participant observation and Chicago school (sociology) were used instead. During the 1970s and 1980s qualitative research began to be used in other disciplines, and became a dominant—or at least significant—type of research in the fields of women's studies, disability studies, education studies, social work studies, information studies, management studies, nursing service studies, human service studies, pshychology and others. In the late 1980s and 1990s after a spate of criticisms from the quantitative side, new methods of qualitative research have been designed, to address the problems with reliability and imprecise modes of data analysis.

An intelligent way of differentiating qualitative research from quantitative research is that largely qualitative research is exploratory, while quantitative research is conclusive. Quantitative data is measurable while qualitative data can not be put into a context that can be graphed or displayed as a mathematical term.

Analytic induction refers to a systematic examination of similarities between various social phenomena in order to develop concepts or ideas. Social scientists doing social research use analytic induction to search for those similarities in broad categories and then develop subcategories. For example, social scientist may examine the category of "marijuana users" and then develop subcategories for "uses marijuana for pleasure" and "uses marijuana for health reasons". If no relevant similarities can be identified, then either the data needs to be re-evaluated and the definition of similarities changed, or the category is too wide and heterogeneous and should be narrowed down.

QUANTITATIVE RESEARCH APPROACHES

Quantitative research is the systematic scientific investigation of quantitative properties and phenomena and their relationships. Quantitative research is widely used in both the natural and social sciences, from physics and biology to sociology and journalism. It is also used as a way to research different aspects of music education. The objective of quantitative research is to develop and employ mathematical models, theories and hypotheses pertaining to natural phenomena. The process of measurement is central to quantitative research because it provides the fundamental connection between empirical observation and mathematical expression of quantitative relationships.

Overview and Background

Quantitative research is generally approached using scientific methods which include:

- The generation of models, theories and hypotheses
- The development of instruments and methods for measurement
- Experimental control and manipulation of variables
- Collection of empirical data
- Modelling and analysis of data
- Evaluation of results

Quantitative research is often an iterative process whereby evidence is evaluated, theories and hypotheses are refined, technical advances are made, and so on. Virtually all research in physics is quantitative whereas research in other scientific disciplines, such as taxonomy and anatomy, may involve a combination of quantitative and other analytic approaches and methods.

Measurement in Quantitative Research

Views regarding the role of measurement in quantitative research are somewhat divergent. Measurement is often regarded as being only a means by which observations are expressed numerically in order to investigate causal relations or associations. However, it has been argued that measurement often plays a more important role in quantitative research.

Quantitative Methods

Quantitative methods are research methods dealing with numbers and anything that is measurable. They are therefore to be distinguished from qualitative methods. Counting and measuring are common forms of quantitative methods. The result of the research is a number, or a series of numbers. These are often presented in tables, graphs or other forms of statistics.

In most physical and biological sciences, the use of either quantitative or qualitative methods is uncontroversial, and each is used when appropriate. In the social sciences, particularly in sociology, social anthropology and psychology, the use of one or other type of method has become a matter of controversy and even ideology, with particular schools of thought within each discipline favouring one type of method and pouring scorn on to the other. Advocates of quantitative methods argue that only by using such methods can the social sciences become truly scientific; advocates of qualitative methods argue that quantitative methods tend to obscure the reality of the social phenomena under study because they underestimate or neglect the non-measurable factors, which may be the most important.

The modern tendency (and in reality the majority tendency throughout the history of social science) is to use eclectic approaches. Quantitative methods might be used with a global qualitative frame. Qualitative methods might be used to understand the meaning of the numbers produced by quantitative methods. Using quantitative methods, it is possible to give precise and testable expression to qualitative ideas.

Descriptive research involves collecting data in order to test hypotheses or answer questions regarding the subjects of the study.

In contrast with the qualitative approach the data are numerical. The data are typically collected through a questionnaire, an interview, or through observation.

In descriptive research, the investigator reports the numerical results for one or more variables on the subjects of the study.

Some examples of descriptive research studies mentioned by Gay are:

i. How do second-grade teachers spend their time?

ii. How will citizens of Yorktown vote in the next election?

iii. How do parents feel about a 12-month school year?

Correlational research attempts to determine whether and to what degree, a relationship exists between two or more quantifiable (numerical) variables. However, it is important to remember that just because there is a significant relationship between two variables it does not follow that one variable causes the other. When two variables are correlated we can use the relationship to predict the value on one variable for a subject if we know that subject's value on the other variable. Correlation implies prediction but not causation. The investigator frequently uses the correlation coefficient to report the results of correlational research.

Some examples of correlational research mentioned by Gay are:

i. The relationship between intelligence and self-esteem.

ii. The relationship between anxiety and achievement.

iii. The use of an aptitude test to predict success in an algebra course.

Causal-comparative research attempts to establish cause-effect relationships among the variables of the study. The attempt is to establish that values of the independent variable have a significant effect on the dependent variable. This type of research usually involves group comparisons. The groups in the study make up the values of the independent variable, for example gender (male versus female), preschool attendance versus no preschool attendance, or children with a working mother versus children without a working mother. These could be the independent variables for the sample studies listed below. However, in causal-comparative research the independent variable is not under the experimenter's control, that is, the experimenter cannot randomly assign the subjects to a gender classification (male or female) but has to take the values of the independent variable as they come. The dependent variable in a study is the outcome variable.

Here are some examples of causal-comparative research studies mentioned by Gay.

i. The effect of preschool attendance on social maturity at the end of the first grade.

ii. The effect of having a working mother on school absenteeism.

iii. The effect of sex (gender) on algebra achievement.

Experimental research like causal-comparative research attempts to establish cause-effect relationship among the groups of subjects that make up the independent variable of the study, but in the case of experimental research, the cause (the independent variable) is under the control of the experimenter. That is, the experimenter can randomly assign subjects to the groups that make up the independent variable in the study. In the typical experimental research design the experimenter randomly assigns subjects to the groups or conditions that constitute the independent variable of the study and then measures the effect this group membership has on another variable, i.e., the dependent variable of the study.

The following are some examples of experimental research mentioned by Gay.

i. The comparative effectiveness of personalized instruction versus traditional instruction on computational skill.

ii. The effect of self-paced instruction on self-concept.

iii. The effect of positive reinforcement on attitude towards school.

SELECTION OF A RESEARCH PROBLEM

Some examples of research problems selected arbitrarily are listed as follows:

1. What are the educational characteristics of gifted learning disabled students in the middle school?

2. Computerized drill versus flash cards to help students with learning disablities remember the addition and subtraction facts.

3. Computer animation as a substitute for concrete materials, for learning basic math concepts.
4. The effect on reading comprehension of using media for the presentation of written materials.

CHARACTERISTICS OF A GOOD RESEARCH PROBLEM

i. The problem is "researchable"—it is a problem that can be investigated through the collection and analysis of data.
ii. The problem has theoretical or practical significance.
iii. It is a good problem for you.

COMPONENTS OF A WELL-STATED RESEARCH PROBLEM

A well-stated research problem should include each of the following components.

    i. The variables of interest to the researcher.
    ii. The specific relationship between the variables.
    iii. The type of subjects involved.

At some point we will need to have an operational definition of each of our variables.

FORMULATING AND CLARIFYING A RESEARCH PROBLEM

With these criteria in mind let's go back and clarify the preliminary research topics we mentioned earlier.

    i. What are the educational characteristics of gifted learning disabled students in the middle school? To make this into a research problem we are going to have to list the variables we are interested in. In other words, What specific variables are implied by "educational characteristics?" Pupils are classified as gifted partly on the basis of cognitive ability (tested IQ). Pupils are classified as learning disabled primarily on the basis of a severe discrepancy between ability and school achievement in reading, math, or written language. With this information we are ready to state our research problem.

    ii. IQ, reading achievement, mathematics achievement, and achievement in written language for gifted students with learning disabilities in the middle school.

iii. Computerized drill versus flash cards to help students with learning disabilities remember the addition and subtraction facts. Here we want to compare two types of instruction in mathematics and evaluate the effect that has on a measure of addition and subtraction facts knowledge. We could state our research problem as:

iv. The effect of type of instruction (computer program versus flash cards) on knowledge of addition and subtraction facts for elementary students with learning disabilities.

v. Computer animation as a substitute for concrete materials, for learning basic math concepts. Modern mathematics instruction has stressed the importance of using concrete models (as well as a problem solving approach) in mathematics instruction. A computer with a two dimensional screen is only capable of showing pictures of the objects rather than the objects themselves. However, some investigators have suggested that animation can be used as a way of making an object more like an actual object than a picture of the object. Of course, some kinds of computer simulations, e.g. virtual reality, would be presenting concrete objects for all practical purposes.

vi. One of the ways we could state this research problem is: The effect of type of presentation (concrete manipulables versus animated computer images) on efficiency in learning addition and subtraction facts by first grade students.

vii. The effect on reading comprehension of using media for the presentation of written materials. In this study, we want to compare the effectiveness of reading from a computer screen with reading printed text on reading speed and reading comprehension. We might also be interested in the type of font used and the size of type. This is developing into a fairly complex study with three independent variables (mode of presentation, type of font, and type size) and two dependent variables (reading speed and reading comprehension). Even though this is a complex study, once we have identified the independent and dependent variables it is fairly easy to state our research problem.

The effect of mode of presentation, type of fonts, and type sizes on reading speed and reading comprehension among middle school students.

4

RESEARCH METHODS

OBSERVATION

Observation is a primary method of collecting data by manual, mechanical, electrical or electronic means. The researcher may or may not have direct contact or communication with the people whose behaviour is being recorded. Observation techniques can be part of **qualitative research** or **quantitative research**. There are six different ways of classifying observation methods:

1. **Participant and nonparticipant** observation, depending on whether the researcher chooses to be part of the situation she/he is studying (e.g. studying social interaction of tour groups by being a tour participant would be participant observation).

2. **Obtrusive and unobtrusive** (or physical trace) observation, depending on whether the subjects being studied can detect the observation (e.g. hidden microphones or cameras observing behaviour and doing garbage audits to determine consumption are examples of unobtrusive observation).

3. Observation in **natural or contrived settings**, whereby the behaviour is observed (usually unobtrusively) when and where it is occurring, while in the contrived setting the situation is recreated to speed up the behaviour.

4. **Disguised and non-disguised** observation, depending on whether the subjects being observed are aware that they are being studied or not. In disguised observation, the researcher may pretend to be someone else, e.g. "just" another tourist participating in the tour group, as opposed to the other tour group members being aware that he/she is a researcher.

5. **Structured and unstructured** observation, which refers to guidelines or a checklist being used for the aspects of the behaviour that are to be recorded; for instance, noting who starts the introductory conversation between two tour group members and what specific words are used by way of introduction.

6. **Direct and indirect** observation, depending on whether the behaviour is being observed as it occurs or after the fact, as in the case of TV viewing, for instance, where choice of program and channel flicking can all be recorded for later analysis.

The data being collected can concern an event or other occurrence rather than people. Although usually thought of as the observation of non-verbal behaviour, this is not necessarily true since comments and/or the exchange between people can also be recorded and would be considered part of this technique, as long as the investigator does not control or in some way manipulate what is being said. For instance, staging a typical sales encounter and recording the responses and reactions by the salesperson would qualify as, observation technique.

One distinct advantage of the observation technique is that it records actual behaviour, not what people say they said/did or believe they will say/do. Indeed, sometimes their actual recorded behaviour can be compared to their statements, to check for the validity of their responses. Especially when dealing with behaviour that might be subject to certain social pressure (for example, people deem themselves to be tolerant when their actual behaviour may be much less so) or conditioned responses (for example, people say they value nutrition, but will pick foods they know to be fatty or sweet), the observation technique can provide greater insight than an actual **survey technique.**

On the other hand, the observation technique does not provide us with any insight into what the person may be thinking or what might motivate a given behaviour/comment. This type of information can only be obtained by asking people directly or indirectly.

When people are being observed, whether they are aware of it or not, ethical issues arise that must be considered by the researcher. Particularly with advances in technology, cameras and microphones have made it possible to gather a significant amount of information about verbal and non-verbal behaviour of customers as well as employees that might easily be considered to be an invasion of privacy or abusive, particularly if the subject is unaware of being observed, yet the information is used to make decisions that impact him/her.

INTERVIEW

The interview method of research, typically, involves a face-to-face meeting in which a researcher (interviewer) asks an individual a series of questions.

What do I need to consider when doing interviews?

- Prepare your interview questions in advance, and share them with the participant(s).
- Tape or videotape the interview.
- Do not be afraid to ask questions during the interview, even if you did not have them listed before the interview.
- After the interview, you will need to transcribe (copy) exactly what was said during the interview. This can be a very slow and time-consuming process, but it is critical that you copy exactly what was said.
- After you have copied out the interview, replay the interview again and compare it to your notes. Make any corrections necessary.
- Share the written copy of the interview with the participant to make sure that they agree with, and affirm, the contents of the interview.

What are the issues, or concerns in conducting interviews?

- Completeness—Did you record the interview and transcribe the interview exactly as recorded?
- Accuracy—Did you miss anything? Did you record it in written form exactly as was said by the participant?
- Bias—Did you "add" to what you observed by presuming or assuming something that was not stated directly by the participant?
- Clarity—Would someone else who had not interviewed the participant be able to get a clear, correct picture of what was discussed by reading your notes?

◎ Confidentiality—Did you ask permission for the interview, and is the participant aware of the purpose and intended audience of the interview?

What materials will I need?

◎ Journal, note paper, writing materials
◎ Tape recorder, videotape recorder
◎ List of interview questions prepared beforehand.

SURVEY

Conducting research using a survey involves going out and asking questions about the phenomenon of interest.

What do I need to consider when doing surveys?

◎ Prepare your survey questions in advance, and share them with your teacher so that they can be checked for accuracy and correctness.

◎ If you are needing additional respondent information such as age, occupation or gender, be sure to include those questions on the survey.

◎ Who is your sample? Will you be getting the kind of information you need from the people you are questioning?

◎ Is your sample size large enough? What sample size is appropriate?

◎ Is your sample representative of the general population? Does it represent a balance between male and female? Do you need to sample a particular age group or will a general survey be alright?

◎ When will you be conducting your survey? How long will you be surveying? Will time lapse make a difference?

◎ How will the respondents answer? Will they record their responses on separate sheets of paper or will you ask them the questions and then record their responses?

◎ How will you present your findings? Will you convert the answers to a percentage? Will you be constructing a bar graph or a pie chart?

◎ If you are considering other factors such as age or gender, then you will need to go through the responses again after you have completed the survey, based on those criteria.

What are the issues, or concerns in conducting surveys?

- ◎   Do the survey questions address the research question?
- ◎   Are the survey questions clear?
- ◎   Is the sample size of respondents large enough?
- ◎   Have I surveyed a representative sample?
- ◎   Apart from factors such as age, occupation and gender, are the respondents anonymous?
- ◎   Is time a factor? How long should I survey? If I surveyed over a longer time period would that affect the results?
- ◎   Are there other sources against which I could compare the survey results? For example, taking the pulse or Reader's Digest surveys?

What materials will I need?

- ◎   Survey questions, response forms
- ◎   Access to online resources.

Types of Surveys

Data are usually collected through the use of questionnaires, although sometimes researchers directly interview subjects. Surveys can use qualitative (e.g. ask open-ended questions) or quantitative (e.g. use forced-choice questions) measures. There are two basic types of surveys: cross-sectional surveys and longitudinal surveys. Much of the following information was taken from the book *Survey Research Methods*, by Earl, R. Babbie.

Cross-sectional surveys   Cross-sectional surveys are used to gather information on a population at a single point in time. An example of a cross-sectional survey would be a questionnaire that collects data on how parents feel about Internet filtering, as of March 1999. A different cross-sectional survey questionnaire might try to determine the relationship between two factors, like religiousness of parents and views on Internet filtering.

Longitudinal surveys   Longitudinal surveys gather data over a period of time. The researcher may then analyse changes in the population and attempt to describe and/or explain them. The three main types of longitudinal surveys are trend studies, cohort studies, and panel studies.

Trend studies   Trend studies focus on a particular population, which is sampled and scrutinized repeatedly. While samples are of the same population, they are typically not composed of the same people. Trend studies, since they may be conducted over a long period of time, do not have to be conducted by just one researcher or research project. A researcher may combine data from several studies of the same population in order to show a trend. An example of a trend study would be a yearly survey of librarians asking about the percentage of reference questions answered using the Internet.

Cohort studies   Cohort studies also focus on a particular population, sampled and studied more than once. But cohort studies have a different focus. For example, a sample of 1999 graduates of GSLIS at the University of Texas could be questioned regarding their attitudes toward paraprofessionals in libraries. Five years later, the researcher could question another sample of 1999 graduates, and study any changes in attitude. A cohort study would sample the same class, every time. If the researcher studied the class of 2004 five years later, it would be a trend study, not a cohort study.

Panel studies   Panel studies allow the researcher to find out why changes in the population are occurring, since they use the same sample of people every time. That sample is called a panel. A researcher could, for example, select a sample of students, and ask them questions on their library usage. Every year thereafter, the researcher would contact the same people, and ask them similar questions, and ask them the reasons for any changes in their habits. Panel studies, while they can yield extremely specific and useful explanations, can be difficult to conduct. They tend to be expensive, take a lot of time, and suffer from high attrition rates. **Attrition** is what occurs when people drop out of the study.

Instrument design   One criticism of library surveys is that they are often poorly designed and administered, resulting in data that is not very accurate, but that is energetically quoted and used to make important decisions. Surveys should be just as rigourously designed and administered as any other research method. Meyer (1998) has identified five preliminary steps that should be taken when embarking upon any research project: 1) choose a topic, 2) review the literature, 3) determine the research question, 4) develop a hypothesis, and 5) operationalization (i.e., figure out how to accurately measure the factors you wish to measure). For research using surveys, two additional considerations are of prime importance: representative sampling and question design.

REPRESENTATIVE SAMPLING

A sample is **representative** when it is an accurate proportional representation of the population under study. If you want to study the attitudes of UG students regarding library services, it would not be enough to interview every 100th person who walked into the library. That technique would only measure the attitudes of UG students who use the library, not those who do not. In addition, it would only measure the attitudes of UG students who happened to use the library during the time you were collecting data. Therefore, the sample would not be very representative of UG students in general. In order to be a truly representative sample, every student at UG would have to have had an equal chance of being chosen to participate in the survey. This is called **randomization**.

If you stood in front of the student union and walked up to students, asking them questions, you still would not have a random sample. You would only be questioning students who happened to come to campus that day, and further, those that happened to walk past the student union. Those students who never walk that way would have had no chance of being questioned. In addition, you might unintentionally be biased as to who you question. You might unconsciously choose not to question students who look preoccupied or busy, or students who do not look like friendly people. This would invalidate your results, since your sample would not be randomly selected.

If you took a list of UG students, uploaded it onto a computer, then instructed the computer to randomly generate a list of 2 per cent of all UG students, then your sample still might not be representative. What if, purely by chance, the computer did not include the correct proportion of seniors, or honours students, or graduate students? In order to further ensure that the sample is truly representative of the population, you might want to use a sampling technique called **stratification**. In order to stratify a population, you need to decide what sub-categories of the population might be statistically significant. For instance, graduate students as a group probably have different opinions than undergraduates regarding library usage, so they should be recognized as separate strata of the population. Once you have a list of the different strata, along with their respective percentages, you could instruct the computer to again randomly select students, this time taking care that a certain percentage are graduate students, a certain percentage are honours students, and a certain percentage are seniors. You would then come up with a more truly representative sample.

SAMPLING AND SAMPLING METHODS

Sampling is that part of statistical practice concerned with the selection of individual observations intended to yield some knowledge about a population of concern, especially for the purposes of statistical inference. Each observation measures one or more properties (weight, location, etc.) of an observable entity enumerated to distinguish objects or individuals.

The sampling process consists of seven simple stages:

1. Defining the population of concern
2. Specifying a sampling frame, a set of items or events possible to measure
3. Specifying a sampling method for selecting items or events from the frame
4. Determining the sample size
5. Implementing the sampling plan
6. Sampling and data collection
7. Reviewing the sampling process

Population and Sampling Frame

Typically, we seek to take action on some population. It is possible to identify and measure every single item in the population. However, in the more general case this is not possible. As a remedy, we seek a **sampling frame** which has the property that we can identify every single element and include any in our sample. The sampling frame must be representative of the population and this is a question outside the scope of statistical theory demanding the judgement of experts in the particular subject matter being studied.

To the scientist, however, representative sampling is the only justified procedure for choosing individual objects for use as the basis of generalization, and is therefore usually the only acceptable basis for ascertaining truth. Having established the frame, there are a number of ways for organizing it to improve efficiency and effectiveness.

Within any of the types of frame identified above, a variety of sampling methods can be employed, individually or in combination.

Types of Sampling

Sampling methods are divided into two types:

1. Probability sampling
2. Non-probability sampling

A probability sample is one in which each member of the population has an equal chance of being selected.

In a non-probability sample, some people have a greater, but unknown, chance than others of selection.

Probability sampling There are six main types of probability sampling.

1. Random sampling
2. Random route sampling
3. Matched random sampling
4. Systematic sampling
5. Stratified sampling
6. Multistage cluster sampling

The choice of these depends on the nature of the research problem, the availability of a good sampling frame, money, time, desired level of accuracy in the sample and data collection methods. Each has its advantages and disadvantages.

Random sampling In random sampling, every combination of items from the frame, or stratum, has a known probability of occurring, but these probabilities are not necessarily equal. There are several forms of random sampling. For example, in simple random sampling, each element has an equal probability of being selected. This method is suitable where population is relatively small and where sampling frame is complete and up-to-date.

Another form of random sampling is Bernoulli sampling in which each element has an equal probability of being selected, like in simple random sampling. However, Bernoulli sampling leads to a variable sample size, while during simple random sampling the sample size remains constant. Bernoulli sampling is a special case of Poisson sampling in which each element may have a different probability of being selected.

Random route sampling   This sampling is used in market research surveys, mainly for sampling households, shops, garages and other premises in urban areas. Addresses are selected at random from sampling frame (usually electoral register) as a starting point. The interviewer is then given instructions to identify further addresses by taking alternate left- and right-hand turns at road junctions and calling at every nth address (shop, garage, etc.).The advantages are, saving of time and reduced bias because of random choice of addresses.

Matched random sampling   This is a method of assigning participants to groups in which pairs of participants are first matched on some characteristic and then individually assigned randomly to groups.

Two samples in which the members are clearly paired, are matched explicitly by the researcher. Example is IQ measurements on pairs of identical twins. Those samples in which the same attribute, or variable, are measured twice on each subject, under different circumstances are commonly called **repeated measures**. Examples include the milk yield of cows before and after being fed a particular diet.

Systematic sampling   Selecting (say) every 10th name from the telephone directory and every 10th sample, is an example of systematic sampling. It is a type of probability sampling unless the directory itself is not randomized. First sampling fraction has to be worked out by dividing population size by required sample size, e.g. for a population of 500 and a sample of 100, the sampling fraction is 1/5, i.e., you will select one person out of every five in the population.

Stratified sampling   Where the population embraces a number of distinct categories, the frame can be organized by these categories into separate "strata." A sample is then selected from each "stratum" separately, producing a stratified sample.

The two main reasons for using a stratified sampling design are

1. to ensure that particular groups within a population are adequately represented in the sample, and

2. to improve efficiency by gaining greater control on the composition of the sample.

For example, if we want to ensure that a sample of 5 students from a group of 50 contains both male and female students in the same proportion as in the full population (i.e., the group of 50), we first divide that population into male and female. In this case, there are 22 male students and 28 females. To work out the number of males and females in the sample:

No. of males in sample $= (5/50) \times 22 = 2.2$

No. of females in sample $= (5/50) \times 28 = 2.8$

We obviously cannot interview .2 of a person or .8 of a person, and have to "round" the numbers. Therefore we choose 2 males and 3 females in the sample. These would be selected using simple random or systematic sample methods.

Multistage cluster sampling   Sometimes it is cheaper to "cluster" the sample in some way, e.g. by selecting respondents from certain areas only, or certain time-periods only. Cluster sampling is an example of "two-stage sampling" or "multistage sampling"—in the first stage a sample of areas is chosen; in the second stage a sample of respondents within those areas is selected. Cluster sampling generally increases the variability of sample estimates above that of simple random sampling, depending on how the clusters differ between themselves, as compared with the within-cluster variation.

Non-probability sampling   It is not always possible to undertake a probability method of sampling, such as in random sampling. For example, there is not a complete sampling frame available for certain groups of the population, e.g. the elderly; people who are attending a football match; people who shop in a particular part of town. Advantages of non-probability methods are cheaper, used when sampling frame is not available and when population is so widely dispersed that cluster sampling would not be efficient.

Purposive sampling   A purposive sample is one which is selected by the researcher subjectively. The researcher attempts to obtain samples that appear to him/her to be representative of the population and will usually try to ensure that a range from one extreme to the other is included. It is often used in political polling, e.g. districts are chosen because their pattern has in the past provided good idea of outcomes for the whole electorate.

Quota sampling   In quota sampling, the population is first segmented into mutually exclusive subgroups, just as in stratified sampling. Then judgement is used to select the subjects or units from each segment based on a specified proportion. For example, an interviewer may be told to sample 200 females and 300 males between the age of 45 and 60. It is this second step which makes the technique one of non-probability sampling. In quota sampling the selection of the sample is non-random and may be biased. For example, interviewers might be tempted to interview those who look most helpful.

Convenience sampling   Sometimes called **grab** or **opportunity sampling**, this is the method of choosing items arbitrarily and in an unstructured manner from the frame. A convenience sample is used when you simply stop anybody in the street who is prepared to stop, or when you wander round a business, a shop, a restaurant, a theatre or whatever, asking people you meet whether they will answer your questions. In other words, the sample comprises subjects who are simply available in a convenient way to the researcher. There is no randomness and the likelihood of bias is high. You cannot draw any meaningful conclusions from the results you obtain.

Sample Size

In addition to the purpose of the study and population size, three criteria usually will need to be specified to determine the appropriate sample size:

1. the level of precision,
2. the level of confidence or risk, and
3. the degree of variability in the attributes being measured.

Sample size depends on the methodology selected, degree of accuracy required for the study (how much error can be tolerated), extent to which there is variation in the population with regard to key characteristics of the study, likely response rate (which itself will depend on sampling method selected) and the time and money available.

The level of precision   The **level of precision**, sometimes called **sampling error**, is the range in which the true value of the population is estimated to be. This range is often expressed in percentage points, (e.g. ± 5 per cent). Thus, if a researcher finds that 60% of farmers in the sample have adopted a recommended practice with a precision rate of ± 5%, then he or she can conclude that between 55% and 65% of farmers in the population have adopted the practice.

The confidence level   The **confidence** or **risk level** is based on ideas encompassed under the Central Limit Theorem. The key idea encompassed in the Central Limit Theorem is that when a population is repeatedly sampled, the average value of the attribute obtained by those samples is equal to the true population value. In a normal distribution, approximately 95% of the sample values are within two standard deviations of the true population value (e.g. mean). In other words, this means that, if a 95% confidence level is selected, 95 out of 100 samples will have the true population value within the range of precision specified earlier.

Degree of variability   The third criterion, the **degree of variability** in the attributes being measured refers to the distribution of attributes in the population. The more heterogeneous a population, the larger the sample size required to obtain a given level of precision. The less variable (more homogeneous) a population, the smaller the sample size. Because a proportion of .5 indicates the maximum variability in a population, it is often used in determining a more conservative sample size.

Determining sample size   There are several approaches for determining the sample size. These include

- using a census for small populations,
- imitating a sample size of similar studies,
- using published tables, and
- applying formula to calculate a sample size.

Types of Data

There are two types of random variables: categorical and numerical.

1. Categorical random variables yield responses such as "yes" or "no". Categorical variables can yield more than two possible responses. For example: "Which day of the week are you most likely to wash clothes?"

2. Numerical random variables yield numerical responses, such as your height in centimetres.

There are two types of numerical variables: discrete and continuous. Discrete random variables produce numerical responses from a counting process. An example is, "how many times do you visit the cash machine in a typical month?" Continuous random variables produce responses from a measuring process. Height is an example of a continuous variable because the response takes on a value from an interval.

Sampling Error

In statistics, sampling error is the error caused by observing a sample instead of the whole population. An estimate of a quantity of interest, such as an average or percentage, will generally be subject to sample-to-sample variation. These variations in the possible sample values of a statistic can theoretically be expressed as sampling errors. Sampling error also refers more broadly to this phenomenon of random sampling variation. The likely size of the sampling error can generally be controlled by taking a large-enough random sample from the population. Error will, therefore, be reduced as the sample size is increased, so that, if a census is performed (a 100 per cent sample is a census), by definition there will be no sampling error. If the observations are collected from a random sample, statistical theory provides probabilistic estimates of the likely size of the sampling error for a particular statistic or estimator. These are often expressed in terms of its standard error.

Experimental Error

Error (or uncertainty) is defined as the difference between a measured or estimated value for a quantity and its true value, and is inherent in all measurements. Experimental error is always with us; it is in the nature of scientific measurement that **uncertainty** is associated with every quantitative result. This may be due to inherent limitations in the measuring equipment, or of the measuring techniques, or perhaps the experience and skill of the experimenter.

Question Design

It is important to design questions very carefully. A poorly designed questionnaire renders results meaningless. There are many factors to consider. The following pointers should be considered while designing a questionnaire:

- Make items clear (don't assume the person you are questioning knows the terms you are using).
- Avoid double-barrelled questions (make sure the question asks only one clear thing).
- Respondent must be competent to answer (don't ask questions that the respondent won't accurately be able to answer).
- Questions should be relevant (do not ask questions on topics that respondents do not care about or haven't thought about).
- Short items are best (so that they may be read, understood, and answered quickly).
- Avoid negative items (if you ask whether librarians should not be paid more, it will confuse respondents).
- Avoid biased items and terms (be sensitive to the effect of your wording on respondents).

Busha and Harter provide the following list of 10 hints:

1. Unless the nature of a survey definitely warrants their usage, avoid slang, jargon, and technical terms.
2. Whenever possible, develop consistent response methods.
3. Make questions as impersonal as possible.
4. Do not bias later responses by the wording used in earlier questions.
5. As an ordinary rule, sequence questions from the general to the specific.
6. If closed questions are employed, try to develop exhaustive and mutually exclusive response alternatives.
7. As far as possible, place questions with similar content together in the survey instrument.
8. Make the questions as easy to answer as possible.
9. When unique and unusual terms need to be defined in questionnaire items, use very clear definitions.
10. Use an attractive questionnaire format that conveys a professional image.

As may be seen, designing good questions is much more difficult than it seems. One effective way of making sure that questions measure what they are supposed to measure is to test them out first, using small focus groups.

SURVEY METHODOLOGY AND DESIGN

Introduction

Everyone has had experience answering surveys, and it is usually a simple and straightforward procedure. Many people conclude from this experience that writing and administering their own survey will be a simple and straightforward matter as well. However, in actuality, the intuitions people form about survey research from their own experience are often incorrect. To test your own intuitions, you might consider how you would answer the following true/false questions:

T/F 1. Determining the opinions of the population of a city of 10,000,000 people requires a much larger sample than an opinion survey of a city of 100,000 people.

T/F 2. Randomly choosing names from a telephone directory is the best way to choose a sample for a telephone survey.

T/F 3. Survey questions should appear in random order.

T/F 4. Posting a survey on a website is a good way to reach large numbers of people and to increase sample size.

T/F 5. If people from the chosen first survey sample fail to respond, a second sample should be chosen to increase the number of respondents.

An understanding of the answers to these and many other questions is essential for conducting scientific surveys that yield accurate, unbiased, and generalizable results. Yet it is a rare person indeed who can explain the reasoning behind such questions correctly without having made an effort to study survey design and methodology. In fact, there really is little reason to expect success in survey research without formal study of the topic. Basic social science research methods are often more complicated, more difficult to learn, and more counterintuitive compared with basic methods in other sciences since the subjects of study, human beings, are more complicated. Atoms and chemicals, for example, don't try to figure out the goals of your research, don't have a bad day, and don't change their minds from one moment to the next!

Social scientists, by conducting countless studies and experiments over the past several decades, now have a good understanding of how to conduct a survey. From such obviously important questions as how to select a random sample to seemingly trivial details such as whether it is better to include a preprinted business reply envelope or a stamped envelope for people to use to return mail surveys, the answers are available in the published academic literature and in textbooks. There are even excellent books on the subject written especially for beginning researchers.

Thus it is now possible for students who have never conducted a survey before to learn about and implement the basic principles of scientific survey design and methodology as part of their IQP. The goal of the present work is to introduce you to these basic principles and to describe where you can go to learn more.

Questionnaire Design

Constructing valid, reliable, and unbiased questions is necessary but not sufficient for creating a good questionnaire: how the questions are organized and presented also deserves careful consideration. The look and feel of a questionnaire serves as an important cue to respondents as they think about how to react to a request to answer a survey. If it is apparent within the first minute or two that the survey is important and easy to complete, people are highly likely to participate; if instead they are not given compelling reasons to take the time away from other activities to answer the survey or if the questions appear to be too difficult, a lot of people will toss the questionnaire into the trash bin or put it on the bottom of their to-do list, resulting in a low response rate. If it is apparent from examining the survey that the researchers put in a lot of time and effort to produce a professional-looking and carefully crafted document, people will likely respond with carefully considered, honest answers; if instead, the survey seems to be poorly organized or contains typographical or other careless errors, respondents will be equally as careless when answering the survey.

Thus questionnaire designers should take several steps to ensure that their instruments make a good impression on potential respondents and to encourage people to respond conscientiously, including the following:

1. Mail surveys should be accompanied by a "cover letter" that briefly introduces the study and explains why it is important and useful. The cover letter should also include three messages that are known to be important for encouraging people to respond:

    i. a promise that the respondent's answers will be kept confidential;

    ii. a statement that describes why their responses, specifically, are necessary for the success of the study; and

    iii. an accurate estimate of the time it will take to complete the survey (which should generally be no more than 10–15 minutes).

2. A good survey is not a random series of questions but is organized. Questions on related topics should be grouped together into sections and placed under descriptive headings. The sections should appear in order from most to least important or most to least closely related to the central topic of the survey.

3. Questionnaires should contain more than just questions. Introductions and transitional statements that briefly explain to respondents what kind of questions they are going to get and why they are important to include in a survey because people find questions much easier to answer when the organization of the survey is made apparent to them.

4. The first few questions on a survey should be carefully chosen, since they must serve to grab respondents' attention and help motivate them to continue to fill out the survey. It is best to begin with a few questions that are easy to answer and that address the most important, central issue of the survey.

5. The format and presentation of the questionnaire must be designed to make it easy to complete without error. Thus, for example, the typeface, type size, and spacing should be easily readable by anyone (some respondents may have poor eyesight!) and the printing should be of high quality. A particularly important feature of good questionnaires is standardization, that is, when possible, different questions should be presented in the same format in order to reduce the time and effort required from the respondent.

CASE STUDY

A case study is an intensive study of one individual. Typically, the case study may involve interviews, observation, experiments and tests.

What do I need to consider when doing case studies?

- Prepare your research questions in advance. What kinds of information do you want to know?
- Consider many different forms of information sources: online websites, paper-based sources such as encyclopaedias, journals, magazines, newspapers, etc.
- If the case study is of a person who can be interviewed, review the following:
  - Prepare your interview questions in advance, and share them with the participant(s).
  - Tape, or videotape the interview.
  - Do not be afraid to ask questions if they arise during the interview, even if you did not have them listed before the interview.
  - After the interview, you will need to transcribe (copy) exactly what was said during the interview. This can be a very slow, and time-consuming process, but it is critical that you copy exactly what was said.
  - After you have copied out the interview, replay the interview again and compare it to your notes. Make any corrections necessary.
  - Share the written copy of the interview with the participants to make sure that they agree with, and affirm the contents of the interview.
- Case studies may also include observational research, experiments and tests. Consider what other types of research are appropriate.

What are the issues, or concerns in conducting case studies?

- Completeness of information recorded is critical to gain a complete understanding of the accuracy of the case study. Have I checked every conceivable resource for information?

◎ Because of the variety of information sources, be sure that you have reviewed all of the issues or concerns for each of the research types.

◎ Guard against bias. Did I "add" to what I observed by presuming or assuming something that was not written about, spoken by or observed of the person?

◎ Would someone else who had not studied the participant be able to get a clear, correct picture of what was discussed by reading your report?

◎ Ensure confidentiality. Be sure you have asked the participants for their permission to be studied, and that they are aware of the purpose and intended audience of the case study report.

What materials will I need?

◎ Journal, note paper, writing materials

◎ Tape recorder, videotape recorder

◎ List of interview questions prepared beforehand

◎ Access to online resources

NATURALISTIC OBSERVATION

In naturalistic observational research the observer does not intervene at all. The researcher is invisible and works hard not to interrupt the natural dynamics of the situation being investigated.

What do I need to consider when doing observational research?

◎ Try to be "invisible", do not get involved in the dynamics of the situation.

◎ Use all of your senses, not just your sense of vision. Record the sounds, smells and tastes (if applicable).

◎ Record your impressions and feelings. How do you feel while observing? Were you frightened, surprised, anxious, amused, excited? Relate what you were feeling to what you were observing.

◎ Record the context of the situation: place, time, participants, number of participants, gender of participants, etc.

◎ Record what you were thinking during the observation. Did the situation remind you of something similar? Had you experienced something similar. What do you think the participants were thinking about while you were observing?

◎ Record all of your information in a journal. Use shorthand or abbreviations if necessary.

What are the issues, or concerns in conducting observational research?

◎ Completeness of information recorded is critical to gain a complete understanding of the dynamics of the situation.

◎ Accuracy of the information recorded is crucial. Did you miss anything? Did you record it exactly as you observed it?

◎ Avoid bias. Did I "add" to what I observed by presuming or assuming something that did not exist?

◎ Would someone else who had not observed the same thing be able to get a clear, correct picture of what you observed by reading your notes?

◎ Respect confidentiality. Be sure not to name people or places in your presentation of the information. You have not asked participants for their permission to conduct research, and so therefore they have the right to remain anonymous.

◎ Refer to the general situation; for example, a school playground, an urban mall, a farm, a family gathering, etc. Videotaping, audiotaping or taking photographs of the situation is infringing on the participant's rights to privacy. Use only your written notes.

What materials will I need?

◎ Journal
◎ Note paper
◎ Writing materials

EXPERIMENTS

Experimental researchers manipulate variables, randomly assign participants to various conditions and seek to control other influences.

What do I need to consider when doing experiments?

◎ Experimental research in psychology involves defining a research problem, describing a hypothesis, describing the process to be followed, gathering data, analysing the data, reporting the findings, and stating conclusions in relation to the hypothesis.

◎ Prepare your experiment in advance. Practice your procedure. Be sure that you have all of the materials necessary to conduct the experiment.

◎ Seek permission to conduct research. Does the participant(s) know that you are conducting an experiment? Do you have his/her written permission?

◎ Is the experiment safe? Will the participants suffer any physical, mental or emotional anguish? If so, do not perform the experiment.

◎ Discuss the experiment with your teacher. Your teacher will be able to guide and assist you to ensure that the experimental procedure is correct, complete and valid.

◎ Audiotape or videotape the experiment. This will help in data analysis, as well as provide some additional information that may be of value when interpreting the data.

What are the issues, or concerns in conducting experiments?

◎ *Ethics of the experiment*   Is the experiment appropriate? Is the experiment safe?

◎ *Bias*   Did I "add" to the test results by presuming or assuming something that was not written about, spoken by or observed during the experiment?

◎ *Confidentiality*   Be sure you have asked participants for their permission to be studied, and that they are aware of the purpose and intended audience of the results of the experiment.

What materials will I need?

◎ Journal, note paper, writing materials

◎ Tape recorder, videotape recorder

◎ List of materials needed, prepared beforehand

◎ Written permission forms.

TOPICAL METHOD OF RESEARCH

A topical research project involves the acquisition, synthesis, organization and presentation of information.

What do I need to consider when doing topical research?

◎ Prepare your research questions in advance: What kinds of information do you want to know?

◎ Consider many different forms of information sources: online websites, paper-based sources such as encyclopaedias, journals, magazines, newspapers, etc.

◎ If the topical research involves a person who can be interviewed, review the following:

  ◎ Prepare your interview questions in advance, and share them with the participant(s).

  ◎ Tape or videotape the interview.

  ◎ Do not be afraid to ask questions if they arise during the interview, even if you did not have them listed before the interview.

  ◎ After the interview, you will need to transcribe (copy) exactly what was said during the interview. This can be a very slow, and time-consuming process, but it is critical that you copy exactly what was said.

  ◎ After you have copied out the interview, replay the interview again and compare it to your notes. Make any corrections necessary.

  ◎ Share the written copy of the interview with the participant to make sure that he/she agrees with, and affirms the contents of the interview.

◎ Topical research studies may also include observational research, experiments and tests. Consider what other types of research are appropriate.

What are the issues, or concerns in conducting topical research studies?

◎ Completeness of information recorded is critical to gain a complete understanding of the topic. Have I checked every conceivable resource for information?

◎ Because of the variety of information sources, be sure that you have reviewed all of the issues or concerns for each of the research types.

◎ Avoid bias. Did I "add" to what I observed by presuming or assuming something that was not written about, spoken by or observed during the research?

◎ Would someone else who had not researched the topic be able to get a clear, correct picture of what the topic was all about by reading your report?

◎ Ensure confidentiality. If you have interviewed or studied individuals connected with the topic, be sure you have asked for their permission to be studied. Ensure that they are aware of the purpose and intended audience of your study.

What materials will I need?

◎ Journal, note paper, writing materials

◎ Tape recorder, videotape recorder

◎ List of interview questions prepared beforehand

◎ Access to online resources.

The survey is a non-experimental, descriptive research method. Surveys can be useful when a researcher wants to collect data on phenomena that cannot be directly observed (such as opinions on library services). Surveys are used extensively in library and information science to assess attitudes and characteristics of a wide range of subjects, from the quality of user-system interfaces to library user reading habits. In a survey, researchers **sample a population**. Busha and Harter (1980) state that "a population is any set of persons or objects that possesses at least one common characteristic." Examples of populations that might be studied are 1) all 1999 graduates of a particular university or 2) all the users of that university's general libraries. Since populations can be quite large, researchers directly question only a sample (i.e., a small proportion) of the population.

5

DESIGNING RESEARCH

All medical research is carried out in relation to one or more objectives, which should focus the plan or design of the research. In some cases there is a clear best way to proceed, but more often there is a choice of reasonable ways of designing a study. The statistical aspects of design relate mainly to the structure of the study and all aspects of the collection of data, including the choice of measurements to make and their frequency.

Research can be crudely divided into observational and experimental studies. In observational studies we collect information about one or more groups of subjects, but do nothing to affect them. Observational studies can be prospective, where subjects are recruited and data are collected about subsequent events, or retrospective, where information is collected about past events. Observational studies include censuses, surveys, case-control studies and cohort studies.

Experimental studies are those in which the researcher affects (controls) what happens to all or some of the individuals. Similar problems arise in studies of humans, animals and laboratory samples, although the emphasis is on clinical studies.

Most studies aim to answer fairly simple questions, but it does not necessarily follow that they require fairly simple designs. The key point is to tailor the research design to the study objective(s). Without adequate planning, the researcher cannot expect to be able to make meaningful conclusions.

In some of the researches we wish to extrapolate the results from a study to the population in general. There are two aspects that require particular attention in this respect. First, the sample(s) studies should be representative of the population(s) of interest; this applies especially to observational studies. Secondly, groups being compared should be as alike as possible apart from the features of direct interest; this applies particularly in experimental studies, such as clinical trials, but is also relevant in many observational studies, such as case-control studies.

CATEGORIES OF RESEARCH DESIGN

Research designs can be classified in several ways, some of which are:

1. observational or experimental,
2. prospective or retrospective, and
3. longitudinal or cross-sectional.

The first classification relates to the purpose of the study, while the others describe the way in which the data are collected.

Observational or Experimental Study

In an observational study, the researcher collects information on the attributes or measurements of interest, but does not influence events. An example would be a study to discover the prevalence of hearing difficulties in small children. Observational studies include surveys and most epidemiological studies. By contrast, in an experimental study, the researcher deliberately influences events and investigates the effects of the intervention. Experimental studies include clinical trials and many animal laboratory studies. In general, stronger inferences can be made from experimental studies than from observational studies. Experimental studies are usually carried out to make comparisons between groups; observational studies may also be comparative, but they are often essentially descriptive.

Prospective or Retrospective Study

There is a clear distinction between prospective studies, in which data are collected forwards in time from the start of the study, and retrospective studies, in which data refer to past events and may be acquired from existing sources, such as hospital notes, or by interview. Experiments are prospective, but observational studies may be prospective or retrospective. Of course, retrospective data can be obtained to compare different treatments, for example different types of mastectomy, but such a study would not be an experiment as it was not a pre-specified study performed under standardized conditions. Retrospective studies include case-control studies.

Longitudinal or Cross-Sectional Study

Longitudinal studies are those which investigate changes over time, possibly in relation to an intervention. Observations are taken on more than one occasion, although they may not all be used in the analysis. Clinical trials are longitudinal because we are interested in the effect of treatment commencing at one time point on outcome at a later time. Cross-sectional studies are those in which individuals are observed only once. Most surveys are cross-sectional, as are studies to construct reference ranges. Observational studies may be longitudinal or cross-sectional, but experiments are usually longitudinal.

There is a clear distinction between experimental studies which are nearly all prospective and longitudinal, and observational studies which can be either retrospective or prospective and also either cross-sectional or longitudinal.

CONTROLS

Whatever the experiment, it is essential to have a comparison, or control, group to which the experimental procedure is not applied. It is not usually scientifically or ethically acceptable to say "Let's try this new treatment on some patients and see what happens". It is far better to have a control group who are treated normally, against which comparisons can be made. If we wish to evaluate the benefits of mothers counting foetal movements in pregnancy we should have a concurrent control group of mothers who do not count movements. This is a key component of the evaluation of new therapies or procedures in medicine.

Controls are also advisable in observational studies. If we ask users of visual display terminals (VDTs) if they get eye strain or backache, we should also ask the same questions to a group of comparable employees who do not use VDTs.

In each case the presence of the control group strengthens the inferences that may be made from the results of the study. However, the choice of suitable controls in observational studies is not easy.

SOURCES OF VARIATION

Variability in behaviour or response to some stimulus, be it tobacco or an antibiotic drug, is the norm. Some sources of variability may be known, or suspected, but much remains unexplained. For example, we know several variables that affect birth weight, such as length of gestation, fetal sex, parity, maternal smoking, height above sea level, and so on, but statistical models incorporating such information explain only about a quarter of the variability in birth weight. While there are undoubtedly other factors not yet identified that contribute to the variability, it is most unlikely that many important factors remain unidentified. The bulk of the observed variability must therefore be considered unexplainable, which we call random variation. There is considerable random variation in most clinical measurements. For some, such as body temperature, there is relatively little variation, but for others, such as birth weight, blood pressure, or many serum constituents, there is enormous variation. When we are designing a study to compare groups with respect to levels of some clinical measurement, this natural variability must be borne in mind.

Further, individuals will exhibit similar variation in other characteristics not directly being studied but which might affect the variables of interest. Many of the principles of experimental design are aimed at trying to control variation that we are not interested in, so that we can focus our attention on the variability that we are interested in. Two consequences of this general variability relevant to the design of studies are:

1.   Care is needed to make samples representative of the population.
2.   In comparative studies, care is needed in making groups similar with respect to known sources of variation.

In addition, we need to bear in mind that when the measurement of interest is highly variable, large samples are needed to get reliable results.

DESIGN OF EXPERIMENTS

An experiment should be designed to answer the question of interest as simply and clearly as possible. It is important to consider the way the data will be analysed when designing an experiment as this can save complications later. The following important aspects are to be considered when designing an experiment.

Bias

Any study, whether experimental or observational, will be set up to answer one or more specific questions. The reliability of the results, and thus the interpretation of the findings, is crucial. An experiment provides the best opportunity to get at the truth, but there are several precautions that should be taken to ensure that the results are not biased. For example, in a comparative experiment, such as the arm comparison study, it is important that the groups of observations being compared are comparable in all aspects other than that being manipulated by the experimenter.

Bias can occur through structural deficiencies in a study. For example, if one observer had taken all measurements on the left arm and the other, all on the right arm, the difference between arms would be inseparable from the differences between the observers, an effect called confounding. In fact, that study was carried out expressly to see if there would be confounding when different machines were compared one to an arm. Making sure that the different observer-arm-cuff combinations were used equally in the 1st, 2nd, 3rd and 4th orders is another example of avoiding bias.

Randomization

An important possible source of bias is the way in which subjects vary in features that are not part of the design. For example, if we had measured blood pressure in the left arm only in one group of patients and in the right arm only in another group, then the average difference observed between left and right arms could be affected by differences between the groups with respect to any variable related to blood pressure, such as age. Clearly it is better to use both arms in the same patients, but in most studies the procedures or treatments cannot be given to the same individuals. The usual approach here is to allocate treatments to patients at random. Random allocation is one of the fundamental principles of experimental design. Another device is to find pairs of subjects with closely similar characteristics and allocate treatments to the matched pair at random.

Random allocation   This refers to allocation of treatments at random to reduce bias in result. There are two main reasons for using randomization. The first reason is to prevent bias. As noted earlier, we want to compare treatments between groups which do not differ in any systemic way. If subjects receive treatments chosen by the investigator there is the likelihood of bias arising—usually subconsciously but occasionally intentionally. We can avoid this possibility by allocating treatments to subjects at random. Bias can also arise through unknown effects. For example, when two or more treatments (or experimental conditions) are used for each subject, it is advisable to randomize the order in which they are applied to each subject in case there is any unknown bias associated with time or the order of measurements. The other reason for randomizing is that statistical theory is based on the idea of random sampling. In a study with random allocation, the differences between treatment groups behave like the differences between random samples.

Simple randomization   By random allocation it is meant that each patient has a known chance, usually an equal chance, of being given each treatment, but the treatment to be given cannot be predicted. Thus, alternately allocating two treatments to a series of patients is not random allocation. The simplest method of random allocation is tossing a coin—heads is treatment A, tails is treatment B. An equivalent method is to use a table of random numbers. In these tables each number occurs equally often, and the ordering is random and so completely unpredictable.

Block randomization   Block (or restricted) randomization is used to keep the number of subjects in the different groups closely balanced at all times. For example, if we consider subjects in blocks of four at a time, there are six ways in which we can allocate treatments so that two subjects get A and two get B:

1. AABB   2. ABAB   3. ABBA
4. BBAA   5. BABA   6. BAAB

If we use combinations of only these six ways of allocating treatments then the numbers in the two groups any time can never differ by more than two, and they will usually be the same or one apart. We choose blocks at random to create the allocation sequence. Randomized blocks can be of any size, but using a multiple of the number of treatments is more logical.

Stratified randomization   While simple randomization removes bias from the allocation procedure, it does not guarantee, for example, that the subjects in each group have similar age distributions. Indeed in small studies, it is highly likely that some chance imbalance will occur, which might complicate the interpretation of results. Even in studies with over 100 subjects there may be some substantial variations by chance, especially for characteristics that are quite rare. In many clinical studies it is known beforehand that subgroups of patients are expected to respond differently to treatment. Here it is advisable to ensure that the subjects receiving each treatment have similar characteristics.

We can use stratified randomization to achieve approximate balance of important characteristics without sacrificing the advantages of randomization. The method is to produce a separate block randomization list for each subgroup. For example, in a study to compare two alternative treatments for breast cancer it would be important to stratify by menopausal status. Two separate lists of random numbers should be obtained, from which two separate piles of sealed envelopes can be prepared, for pre-menopausal and post-menopausal women. It is essential that stratified treatment allocation is based on block randomization within each stratum rather than simple randomization; otherwise there will be no control of balance of treatments within strata, and so the object of stratification will be defeated.

Stratified randomization can be extended to two or more stratifying variables. For example, we might wish to extend the stratification in the breast cancer trial to tumour size and number of positive nodes. We have to produce a separate randomization list for each combination of categories. If we had two tumour size groups and three groups for node involvement as well as menopausal status, then we have $2 \times 3 \times 2 = 12$ strata, which may exceed the limit of what is practical. There is the further problem with multiple strata that some of the combinations of categories may be rare, so that the treatment balance expected from the use of block randomization does not occur.

Blinding

Bias can also occur through subconscious effects. For example, observers' judgements may be affected by knowing the treatment that a subject is getting, or by knowledge of a previous measurement for that subject. The latter problem was avoided in the arm comparison study by the choice of blood pressure measuring machine. The former problem is especially relevant in clinical trials, where it is desirable to keep both patients and assessors in ignorance of the treatment given, a procedure known as blinding.

Replication

For measurements that are highly variable or difficult to measure accurately it may be useful to take more than one measurement on each individual. These replicates can be treated in the analysis as separate observations, which may make the analysis more complicated but gives greater potential to detect effects of interest. This analysis is only valid if the replicates are independent, which is often not the case if the observer knows what measurement they obtained the first time. More often, the average of the replicates is used in the analysis. This latter approach may mirror clinical practice—some "noisy" variables such as blood pressure, peak expiratory flow rate, and ultrasound measurements are usually repeated.

Sample Selection

It is always desirable for the sample in a study to be representative of the population of interest, but this is not as important in experiments as in observational studies. For example, it is unlikely that the choice of the sample for the arm difference study would have affected the results. It is much more important to ensure that the subgroups being compared are as similar as possible. Although in principle representative samples are best obtained by random selection from the population, this ideal is virtually never met in practice. However, the sample should be chosen to be as similar as possible to the relevant population, so it is essential to be able to describe just how the sample was chosen.

Sample Size

Another way of combating variability is to increase the sample size. Larger samples enable us to evaluate effects of interest more precisely. The determination of an appropriate sample size is most common in clinical trials.

Minimization

The only form of allocation that is an acceptable alternative to randomization is minimization, which is a clever method of ensuring excellent balance between the groups for several prognostic factors, even in small samples. It is based on the idea that the next patient to enter the trial is given, with probability greater than 0.5, whichever treatment would minimize the overall imbalance between the groups at that stage of the trial. Often the probability is taken as 1, but a value greater than, say 0.75, should achieve much the same result with the advantages of a random component.

Observational Studies

Many studies are carried out to investigate possible associations between various factors and the development of a particular disease or condition. Examples are studies of the relationship between passive smoking and lung cancer, the use of visual display terminals and miscarriage, and alcohol consumption and suicide. There is no logical difference between comparing the outcome of two groups of patients given alternative treatments and comparing the outcome of groups receiving different exposures. In general, however, areas of epidemiological research are not amenable to being investigated by randomized trials. We cannot randomize individuals to smoke or not to smoke nor to work in particular jobs, and other factors such as age and race are not controllable by the individual. We must use observational studies, to study factors or exposures which cannot be controlled by the investigators.

There are two main types of observational study that are used to investigate causal factors—the case-control study and the cohort study. In a retrospective case-control study a number of subjects with the disease in question (the cases) are identified along with some unaffected subjects (controls). The past history of these groups in relation to exposure(s) of interest is then compared. In contrast, in a prospective cohort study, a group of subjects is identified and followed prospectively, perhaps for many years, and their subsequent medical history recorded. The cohort may be subdivided at the outset into groups with different characteristics, or the study may be used to investigate which subjects go on to develop a particular disease.

The case-control study   In the case-control study, we identify a group of subjects (cases) with the disease condition or condition of interest, say lung cancer, and an unaffected group (controls), and compare their past exposure to one or more factors of interest, such as consumption of carrots. If the cases report greater exposure than the controls we may infer that exposure is causally related to the disease of interest, for example that consumption of carrots affects the risk of developing lung cancer.

The prime advantages of the case-control approach are practical: it is relatively simple, and thus quick and cheap. The case-control design is also valuable when the condition of interest is very rare. The disadvantages of this design are important, and relate to possible biases in the comparison of cases and controls.

The cohort study   The prospective cohort study (or follow-up or longitudinal study) is the method of choice for an observational study, but there are certain difficulties with this design too. The essence of the cohort study is to identify a group of subjects of interest and then follow them up to see what happens. Because of the need to observe unaffected individuals until a fair proportion develop the outcome of interest, cohort studies can take a long time and may thus be very expensive. They are usually unsuitable for studying rare outcomes as it would be necessary to follow a huge number of subjects to get an adequate number of events.

There is usually one particular event of interest, such as death or recurrence of disease, but there may be several. There may be subgroups of subjects identified at the outset whose experience is to be compared, such as smokers and non-smokers or patients with different stages of breast cancer. Alternatively the purpose of the study may be to use the information gained to try to identify those subjects most at risk of developing the outcome of interest. For example, we could follow patients with cirrhosis of the liver, identify those developing carcinoma of the liver over, say, ten years, and compare their characteristics with those who do not get a carcinoma. Because the study is prospective, the nature and quality of the data recording can be carefully controlled.

The cross-sectional study   In a cohort study, subjects with different characteristics are identified and followed to see what happens. By contrast, in a cross-sectional study, all the information is collected at the same time because subjects are only contacted once. Many cross-sectional studies are descriptive, and these are often called surveys. For example, we might ask undergraduates about their alcohol consumption, carry out a survey of the use of alternative medicine in a particular area, or investigate the ability of a particular blood test to give a correct "diagnosis" in patients with certain symptoms. The cross-sectional study does not suffer from many of the difficulties that affect these other designs, such as recall bias and loss to follow-up. It is relatively cheap and easy to carry out.

Choosing a Study Design

The choice between an experiment and an observational study is usually straightforward. If it is possible, both ethically and logistically, to carry out an experiment, then this is the preferred approach. In particular, the evaluation of alternative treatments is best addressed by a randomized controlled trial. A review of papers published in the *New England Journal of Medicine* in 1978–79 found that only 90 of 332 original articles were controlled experiments and the proportion is probably unusually high in that journal. The majority of the remainder were observational studies and most of those were cross-sectional studies.

The choice of the most appropriate design is not easy, as there are many considerations to weigh up. The involvement of a statistician at the planning stage is strongly recommended. In addition to advising on the choice of design, they can give valuable assistance regarding the selection of suitable samples of individuals for study, a problem that must be confronted with any study design but is especially important in observational studies. The statistician can (and should) also advise on the appropriate sample size.

6

DATA COLLECTION

Data collection techniques allow us to **systematically** collect information about our objects of study (people, objects, phenomena) and about the settings in which they occur. In the collection of data we have to be systematic. If data are collected haphazardly, it will be difficult to answer our research questions in a conclusive way.

Various data collection techniques can be used such as:

◎ Using available information

◎ Observing

◎ Interviewing (face-to-face)

◎ Administering written questionnaires

◎ Focus group discussions

◎ Projective techniques, mapping, scaling

USING AVAILABLE INFORMATION

Usually there is a large amount of data that has already been collected by others, although it may not necessarily have been analysed or published. Locating these sources and retrieving the information is a good starting point in any data collection effort.

For example, analysis of the information routinely collected by health facilities can be very useful for identifying problems in certain interventions or in flows of drug supply, or for identifying increases in the incidence of certain diseases.

Analysis of health information system data, census data, unpublished reports and publications in archives and libraries or in offices at the various levels of health and health-related services, may be a study in itself. Usually, however, it forms part of a study in which other data collection techniques are also used.

The advantage of using existing data is that collection is inexpensive. However, it is sometimes difficult to gain access to the records or reports required, and the data may not always be complete and precise enough, or may be too disorganized.

Often used as a data source as the information desired has already been obtained; it cuts down time on the data collection process.

Example   Study that looked at court records to obtain information about conservatorship of elderly patients.

Problems   Issue of privacy—the Privacy Act states that the individual must not be able to be identified.

The data on the records must be obtained in a similar fashion and recorded in such a way that consistency can be obtained.

OBSERVATION

Observation is a technique that involves systematically selecting, watching and recording behaviour and characteristics of living beings, objects or phenomena.

Observation of human behaviour is a much-used data collection technique. It can be undertaken in different ways:

- **Participant observation**   The observer takes part in the situation he or she observes. (For example, a doctor hospitalized with a broken hip, who now observes hospital procedures "from within".)
- **Non-participant observation**   The observer watches the situation, openly or concealed, but does not participate.

Observations can be **open** (e.g. "shadowing" a health worker with his/her permission during routine activities) or **concealed** (e.g. "mystery clients" trying to obtain antibiotics without medical prescription). They may serve different purposes. Observations can give additional, more accurate information on behaviour of people than interviews or questionnaires. They can also check on the information collected through interviews especially on sensitive topics such as alcohol or drug use, or stigmatizing diseases. For example, whether community members share drinks or food with patients suffering from feared diseases (leprosy, TB, AIDS) are essential observations in a study on stigma.

Observations of human behaviour can form part of any type of study, but as they are time-consuming they are most often used in small-scale studies.

Observations can also be made on **objects**. For example, the presence or absence of a latrine and its state of cleanliness may be observed. Here observation would be the major research technique.

If observations are made using a defined scale they may be called **measurements**. Measurements usually require additional tools. For example, in nutritional surveillance, we measure weight and height by using weighing scales and a measuring board. We use thermometers for measuring body temperature.

1. The observations must be consistent with the theoretical framework chosen for the study.
2. The recording of the observations must be systematic and standardized.
3. Controls must be used to keep the environmental influences from skewing the data.
4. Consider the role of the observer. Is he/she concealed or not concealed?
5. Does the observer carry out an intervention or does he just observe?
6. Ethical issues must be considered. Was informed consent used? How would this affect the subject's performance? If the subject is not informed until after the observation is completed (debriefing) has the person's rights been violated?
7. The effect of reactivity (Hawthorne effect).

INTERVIEWING

An interview is a data-collection technique that involves oral questioning of respondents, either individually or as a group. Answers to the questions posed during an interview can be recorded by writing them down (either during the interview itself or immediately after the interview) or by taping the responses, or by a combination of both. Interviews can be conducted with varying degrees of flexibility. The two extremes, high and low degree of flexibility, are described below.

High Degree of Flexibility

When studying sensitive issues such as teenage pregnancy and abortions, for example, the investigator may use a list of topics rather than fixed questions. These may include how teenagers started sexual intercourse, the responsibility girls and their partners take to prevent pregnancy (if at all), and the actions they take in the event of unwanted pregnancies. The investigator should have an additional list of topics ready when the respondent falls silent, (e.g. when asked about abortion methods used, who made the decision and who paid). The sequence of topics should be determined by the flow of discussion. It is often possible to come back to a topic discussed earlier in a later stage of the interview.

The unstructured or loosely structured method of asking questions can be used for interviewing individuals as well as groups of key informants. A flexible method of interviewing is useful if a researcher has as yet little understanding of the problem or situation he is investigating, or if the topic is sensitive. It is frequently applied in exploratory studies. The instrument used may be called an **interview guide** or interview schedule.

Low Degree of Flexibility

Less flexible methods of interviewing are useful when the researcher is relatively knowledgeable about expected answers or when the number of respondents being interviewed is relatively large. Then **questionnaires** may be used with a fixed list of questions in a standard sequence, which have mainly fixed or pre-categorized answers.

For example after a number of observations on the (hygienic) behaviour of women drawing water at a well and some key informant interviews on the use and maintenance of the wells, one may conduct a larger survey on water use and satisfaction with the quantity and quality of the water.

There are different types of interviews like face-to-face interviews, telephone interviews and computer-assisted personal interviews.

Face-to-face interviews have a distinct advantage of enabling the researcher to establish rapport with potential participants and therefore gain their cooperation. These interviews yield highest response rates in survey research. They also allow the researcher to clarify ambiguous answers and when appropriate, seek follow-up information. The disadvantages are that they are impractical when large samples are involved and they are time-consuming and expensive.

Telephone interviews are less time-consuming and less expensive and the researcher has ready access to anyone on the planet who has a telephone. Disadvantages are that the response rate is not as high as the face-to-face interviews, but is considerably higher than the mailed questionnaire. The sample may be biased to the extent that people without phones are part of the population about whom the researcher wants to draw inferences.

Computer-assisted personal interviews (CAPI) is a form of personal interviewing, but instead of completing a questionnaire, the interviewer brings along a laptop or hand-held computer to enter the information directly into the database. This method saves time involved in processing the data, as well as saves the interviewer from carrying around hundreds of questionnaires. However, this type of data collection method can be expensive to set up and requires that interviewers have computer and typing skills.

WRITTEN QUESTIONNAIRES

A written questionnaire (also referred to as self-administered questionnaire) is a data collection tool in which written questions are presented that are to be answered by the respondents in written form.

A written questionnaire can be administered in different ways, such as:

- By sending questionnaires by mail with clear instructions on how to answer the questions and asking for mailed responses;

- By gathering all or part of the respondents in one place at one time, giving oral or written instructions, and letting the respondents fill out the questionnaires; or

- By hand-delivering questionnaires to respondents and collecting them later.

The questions can be either open-ended or closed (with pre-categorized answers).

Paper–pencil questionnaires can be sent to a large number of people and this saves the researcher time and money. People are more truthful while responding to the questionnaires regarding controversial issues in particular due to the fact that their responses are anonymous. But they also have drawbacks. Majority of the people who receive questionnaires do not return them and those who do, might not be representative of the originally selected sample.

Web-based questionnaires—a new and inevitably growing methodology is the use of Internet-based research. This would mean receiving an e-mail on which you would click on an address that would take you to a secure website to fill in a questionnaire. This type of research is often quicker and less detailed. Some disadvantages of this method include the exclusion of people who do not have a computer or are unable to access a computer. Also the validity of such surveys are in question as people might be in a hurry to complete it and so might not give accurate responses.

Questionnaires often make use of checklist and rating scales. These devices help to simplify and quantify people's behaviours and attitudes. A **checklist** is a list of behaviours, characteristics, or other entities that the researcher is looking for. Either the researcher or survey participant simply checks whether each item on the list is observed, present or true or vice versa. A **rating scale** is more useful when a behaviour needs to be evaluated on a continuum. They are also known as Likert scales.

FOCUS GROUP DISCUSSIONS (FGD)

A focus group discussion (FGD) is a group discussion of approximately 6–12 persons guided by a facilitator, during which group members talk freely and spontaneously about a certain topic.

An FGD is a qualitative method. Its purpose is to obtain in-depth information on concepts, perceptions and ideas of a group. An FGD aims to be more than a question–answer interaction. The idea is that group members discuss the topic among themselves, with guidance from the facilitator.

FGD techniques can, for example, be used to:

1. Focus research and develop relevant research hypotheses by exploring in greater depth the problem to be investigated and its possible causes.

   Example   A district health officer had noticed that there were an unusually large number of cases of malnutrition of children/ under age 5 reported from one area in her district. Because she had little idea of why there might be more malnutrition in that area she decided to organize three focus group discussions (one with leaders, one with mothers from the area and one with health staff from the area). She hoped to identify potential causes of the problem through the FGDs and then develop a more intensive study, if necessary.

2. Formulate appropriate questions for more structured, large-scale surveys.

   Example   In planning a study of the incidence of childhood diarrhoea and feeding practices, a focus group discussion showed that in the community under study, children below the age of 1 year were not perceived as having "bouts of diarrhoea" but merely "having loose stools" that were associated with milestones such as sitting up, crawling, and teething. In the questionnaire that was developed after the FGD the concept "diarrhoea" was therefore carefully described, using the community's notions and terms.

3. Help understand and solve unexpected problems in interventions.

   Example   In District X, the recent national (polio) immunization days (NID) showed widely different coverages per village (50–90%) and in a number of villages a marked decrease in coverage was observed compared to last year. Eight FGDs were held with mothers, two in the town, three in rural villages with a marked decrease in NID coverage and three in villages with a high coverage throughout. It appeared that overall, the concept NID had raised confusion. Most people believed that this mass campaign strengthened the children's immunity against *any* (childhood) disease, including malaria and respiratory tract infections. In the villages with a low NID coverage there had been a high incidence of malaria in children immediately after the previous NID campaign and several children died. Mothers therefore believed that the NID campaign was useless.

4. Develop appropriate messages for health education programmes and later evaluate the messages for clarity.

    Example   A rural health clinic wanted to develop a health education programme focused on weaning problems most often encountered by mothers in the surrounding villages and what to do about them. The focus group discussion could be used for **exploring relevant local concepts** as well as for **testing drafts** when developing the messages. The messages should be developed and tested in different socio-economic groups of mothers, as weaning practices may differ with income, means of subsistence and education of the mothers. Also ethnic differences may have to be taken into account.

5. Explore controversial topics.

    Example   Sexual behaviour is a controversial topic in the sense that males and females judge sexual relations and sexuality often from very different perspectives. Sexual education has to take this difference into account. Through FGDs, first with females, then with males, and then with a mixed group to confront both sexes with the different outcomes of the separate discussions it becomes easier to bring these differences in the open. Especially for teenagers, who may have many stereotypes about the other sex or are reluctant to discuss the topic openly (particularly girls), such a "multistage" approach is useful.

Strengths and Limitations

Implementation of FGDs is an iterative process; each focus group discussion builds on the previous one, with a slightly elaborated or better-focused set of themes for discussion. Provided the groups have been well-chosen, in terms of composition and number, FGDs can be a powerful research tool providing valuable spontaneous information in a short period of time and at relatively low cost.

FGD should *not* be used for quantitative purposes, such as the testing of hypotheses or the generalization of findings for larger areas, which would require more elaborate surveys.

However, FGDs can profitably complement such surveys or other, qualitative techniques. Depending on the topic, it may be risky to use FGDs as a single tool. In group discussions, people tend to centre their opinions on the most common ones, on "social norms". In reality, opinions and behaviour may be more diverse. Therefore it is advisable to combine FGDs with at least some key informant and in-depth interviews. Explicitly soliciting other views during FGDs should be routine as well.

In case of very sensitive topics such as sexual behaviour or coping with HIV/AIDS, FGDs may also have their limitations, as group members may hesitate to air their feelings and experiences freely. One possible remedy is the selection of participants who do not know each other (e.g. selection of children from different schools in FGDs about adolescent sexual behaviour), while assuring absolute confidentiality.

It may also help to alternate the FGD with other methods, for example, to precede it by a self-developed role play on sexual behaviour, or to administer a written questionnaire immediately after the FGD with open questions on sexual behaviour in which the participants can anonymously state all their questions and problems. This worked in Tanzania and Nepal.

PROJECTIVE TECHNIQUES

When a researcher uses projective techniques, (s)he asks an informant to react to some kind of visual or verbal stimulus.

For example, an informant may be provided with a rough outline of the body and be asked to draw her or his perception of the conception or onset of an illness.

Another example of a projective technique is the presentation of a hypothetical question or an incomplete sentence or case study to an informant ("story with a gap"). A researcher may ask the informant to complete in writing sentences such as:

— If I were to discover that my neighbour had TB, I would ...;

— If my wife were to propose that I use condoms, I would ...

Or (s)he may ask the informant: Suppose your child suffered from diarrhoea, what would you do?

Such techniques can easily be combined with semi-structured interviews or written questionnaires. They are also very useful in FGDs to get people's opinion on sensitive issues.

MAPPING AND SCALING

Mapping is a valuable technique for visually displaying relationships and resources.

In a water supply project, **for example**, mapping is invaluable. It can be used to present the placement of wells, distance of the homes from the wells, other water systems, etc. It gives researchers a good overview of the physical situation and may help to highlight relationships hitherto unrecognized.

Mapping a community is also very useful and often indispensable as a pre-stage to sampling.

Scaling is a technique that allows researchers through their respondents to categorize certain variables that they would not be able to rank themselves.

For example, they may ask their informant(s) to bring certain types of herbal medicine and ask them to arrange these into piles according to their usefulness. The informants would then be asked to explain the logic of their ranking.

Mapping and scaling may be used as participatory techniques in rapid appraisals or situation analyses. In a separate volume on participatory action research, more such techniques will be presented.

Rapid appraisal techniques and participatory research are approaches often used in health systems research.

DATA COLLECTION TECHNIQUES AND TOOLS

To avoid confusion in the use of terms, the following table points out the distinction between techniques and tools applied in data collection.

Data collection techniques	Data collection tools
Using available information	Checklist; data compilation forms
Observing	Eyes and other senses, pen/paper, watch, scales, microscope, etc.
Interviewing	Interview guide, checklist, questionnaire, tape recorder
Administering written questionnaires	Questionnaire

Advantages and Disadvantages of Various Data Collection Techniques

Techniques	Advantages	Possible constraints
Using available information	Is inexpensive, because data is already there. Permits examination of trends over the past.	Data is not always easily accessible. Ethical issues concerning confidentiality may arise. Information may be imprecise or incomplete.
Observing	Gives more detailed and context-related information. Permits collection of information on facts not mentioned in an interview. Permits tests of reliability of responses to questionnaires.	Ethical issues concerning confidentiality or privacy may arise. Observer bias may occur. (Observer may only notice what interests him or her.) The presence of the data collector can influence the situation observed. Thorough training of research assistants is required.
Interviewing	Is suitable for use with both literates and illiterates. Permits clarification of questions. Has higher response rate than written questionnaires.	The presence of the interviewer can influence responses. Reports of events may be less complete than information gained through observations.

(Contd.)

Techniques	Advantages	Possible constraints
Small-scale flexible interview	Permits collection of in-depth information and exploration of spontaneous remarks by respondents.	The interviewer may inadvertently influence the respondents. Analysis of open-ended data is more difficult and time-consuming.
Large-scale fixed interview	Is easy to analyse.	Important information may be missed because spontaneous remarks by respondents are usually not recorded or explored.
Administering written questionnaires	Is less expensive. Permits anonymity and may result in more honest responses. Does not require research assistants. Eliminates bias due to phrasing questions differently with different respondents.	Cannot be used with illiterate respondents. There is often a low rate of response. Questions may be misunderstood.
Participatory and projective methods	Provide rich data and may have positive spin offs for knowledge and skills by researchers and informants	Require some extra training of researchers.

IMPORTANCE OF COMBINING DIFFERENT DATA COLLECTION TECHNIQUES

When discussing different data collection techniques and their advantages and disadvantages, it becomes clear that they can complement each other. The skilful use of a combination of different techniques can reduce the chance of bias and will give a more comprehensive understanding of the topic under study.

Researchers often use a combination of flexible and less flexible research techniques.

Flexible techniques, such as

◎   loosely structured interviews using open-ended questions,

◎   focus group discussions, and

◎   participant observation.

are also called **qualitative research techniques**. They produce qualitative data that is often recorded in narrative form.

Qualitative research techniques involve the identification and exploration of a number of often mutually related variables that give insight in human behaviour (motivations, opinions, attitudes), in the nature and causes of certain problems and in the consequences of the problems for those affected. "Why", "What" and "How" are important questions.

Qualitative data collection methods play an important role in impact evaluation by providing information useful to understand the processes behind observed results and assessing changes in people's perceptions of their well-being. Furthermore qualitative methods can be used to improve the quality of survey-based quantitative evaluations by helping generate evaluation hypothesis; strengthening the design of survey questionnaires and expanding or clarifying quantitative evaluation findings. These methods are characterized by the following attributes.

◎   They tend to be open-ended and have less structured protocols (i.e., researchers may change the data collection strategy by adding, refining, or dropping techniques or informants).

◎   They rely more heavily on iterative interviews; respondents may be interviewed several times to follow up on a particular issue, clarify concepts or check the reliability of data.

◎   They use triangulation to increase the credibility of their findings (i.e., researchers rely on multiple data collection methods to check the authenticity of their results).

◎   Generally their findings are not generalizable to any specific population, rather each case study produces a single piece of evidence that can be used to seek general patterns among different studies of the same issue.

Regardless of the kinds of data involved, data collection in a qualitative study takes a great deal of time. The researcher needs to record any potentially useful data thoroughly, accurately, and systematically, using field notes, sketches, audiotapes, photographs and other suitable means. The data collection methods must observe the ethical principles of research.

The qualitative methods most commonly used in evaluation can be classified in three broad categories:

1   in-depth interview

2   observation methods

3   document review

Different ways of collecting evaluation data are useful for different purposes, and each has advantages and disadvantages. Various factors will influence your choice of a data collection method: the questions you want to investigate, resources available to you, your timeline, and more.

Structured questionnaires that enable the researcher to quantify pre- or post-categorized answers to questions are an example of **quantitative** research techniques. The answers to questions can be counted and expressed numerically.

Quantitative research techniques are used to quantify the size, distribution, and association of certain variables in a study population. "How many?" "How often?" and "How significant?" are important questions.

Both qualitative and quantitative research techniques are often used within a single study.

The **quantitative data collection methods**, rely on random sampling and structured data collection instruments that fit diverse experiences into predetermined response categories. They produce results that are easy to summarize, compare, and generalize.

Quantitative research is concerned with testing hypotheses derived from theory and/or being able to estimate the size of a phenomenon of interest. Depending on the research question, participants may be randomly assigned to different treatments. If this is not feasible, the researcher may collect data on participant and situational characteristics in order to statistically control their influence on the dependent, or outcome, variable. If the intent is to generalize from the research participants to a larger population, the researcher will employ probability sampling to select participants.

Typical quantitative data-gathering strategies include:

- Experiments/clinical trials.
- Observing and recording well-defined events (e.g. counting the number of patients waiting in emergency at specified times of the day).
- Obtaining relevant data from management information systems.

Administering surveys with close-ended questions (e.g. face-to-face and telephone interviews, questionnaires, etc.).

BIAS IN INFORMATION COLLECTION

Bias in information collection is a distortion in the collected data, so it does not represent reality.

Possible Sources of Bias During Data Collection

1. *Defective instruments* These include:

- Questionnaires with
    - fixed or closed questions on topics about which little is known (often asking the "wrong things");
    - open-ended questions without guidelines on how to ask (or to answer) them;
    - vaguely phrased questions;
    - "leading questions" that cause the respondent to believe one answer would be preferred over another; or
    - questions placed in an illogical order.
- Weighing scales or other measuring equipment that are not standardized.

These sources of bias can be prevented by **carefully planning the data collection process** and by **pre-testing the data collection tools.**

2. *Observer bias* Observer bias can easily occur when conducting observations or utilizing loosely structured group interviews or individual interviews. There is a risk that the data collector will only see or hear things in which (s)he is interested or will miss information that is critical to the research.

Observation protocols and guidelines for conducting loosely structured interviews should be prepared, and training and practice should be provided to data collectors in using both these tools. Moreover it is highly recommended that data collectors **work in pairs** when using flexible research techniques and discuss and interpret the data immediately after collecting it. Another possibility, commonly used by anthropologists, is using a tape recorder and transcribing the tape word by word.

3. *Effect of the interview on the informant*   This is a possible factor in all interview situations. The informant may mistrust the intention of the interview and dodge certain questions or give misleading answers. For example, in a survey on alcoholism you ask school children: "Does your father sometimes get drunk?" Many will probably deny that he does, even if it is true. Such bias can be reduced by adequately introducing the purpose of the study to informants, by phrasing questions on sensitive issues in a positive way, by taking sufficient time for the interview, and by assuring informants that the data collected will be confidential.

It is also important to be careful in the selection of interviewers. In a study soliciting the reasons for the low utilization of local health services, for example, one should not ask health workers from the health centres concerned to interview the population. Their use as interviewers would certainly influence the results of the study.

4. *Information bias*   Sometimes the information itself has weaknesses. Medical records may have many blanks or be unreadable. This tells something about the quality of the data and has to be recorded. For example, in a TB defaulter study the percentage of defaulters with an incomplete or missing address should be calculated.

Another common information bias is due to gaps in people's memory; this is called **memory** or **recall bias**. A mother may not remember all details of her child's last diarrhoea episode and of the treatment she gave two or three months afterwards. For such common diseases it is advisable to limit the period of recall, asking, for example, "Has your child had diarrhoea over the past two weeks?"

ETHICAL CONSIDERATIONS

As we develop our data collection techniques, we need to consider whether our research procedures are likely to cause any physical or emotional harm. Harm may be caused, for example, by:

- violating informants' right to privacy by posing sensitive questions or by gaining access to records which may contain personal data;

- observing the behaviour of informants without their being aware (concealed observation should therefore always be cross-checked or discussed with other researchers with respect to ethical admissibility);

- allowing personal information to be made public which informants would want to be kept private, and

- failing to observe/respect certain cultural values, traditions or taboos valued by your informants.

Several methods for dealing with these issues may be recommended:

- obtaining informed consent before the study or the interview begins;

- not exploring sensitive issues before a good relationship has been established with the informant;

- ensuring the confidentiality of the data obtained; and

- learning enough about the culture of informants to ensure it is respected during the data collection process.

If sensitive questions are asked, for example, about family planning or sexual practices, or about opinions of patients on the health services provided, it may be advisable to omit names and addresses from the questionnaires.

7

DATA ANALYSIS

The data analysis is divided into qualitative and quantitative data analysis.

QUALITATIVE ANALYSIS

This is the process of interpreting data collected during the course of qualitative research.

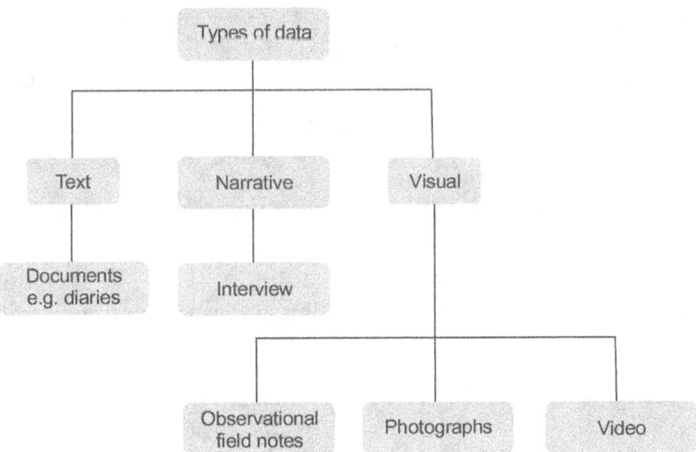

The analysis of the data depends on its type. Within this unit there is further guidance on the analysis of visual data and the analysis of narrative data.

QUANTITATIVE ANALYSIS

This is the process of presenting and interpreting numerical data.

The results section of papers including quantitative data analysis often contain descriptive statistics and inferential statistics.

Descriptive statistics include measures of central tendency (averages—mean, median and mode) and measures of variability about the average (range and standard deviation). These give the reader a picture of the data collected and used in the research project.

Inferential statistics is the outcome of statistical tests, helping deductions to be made from the data collected, to test hypotheses set and relating findings to the sample or population.

Descriptive or Summary Statistics

Quantitative research may well generate masses of data. For example, a comparatively small study that distributes 200 questionnaires with 20 items on each can generate potentially 4000 items of raw data.

To make sense of this data, it needs to be summarized in some way, so that the reader has an idea of the typical values in the data, and how they vary. To do this, researchers use descriptive or summary statistics: they describe or summarize the data, so that the reader can construct a mental picture of the data and the people, events or objects they relate to.

All quantitative studies will have some descriptive statistics, as well as frequency tables, for example, sample size, maximum and minimum values, averages and measures of variation of the data about the average. In many studies this is the first step, prior to more complex inferential analysis.

The two main types of descriptive statistics encountered in research papers are measures of central tendency (averages) and measures of dispersion.

Inferential Statistics

A set of measurements can almost always be regarded as measurements on a sample of items from a population of these items, as it is usually impractical or impossible to measure every item in the population.

Thus we have to make inferences about the population from the sample.

This can be true only if the sample is representative of the population, and even then the sample is very unlikely to reflect the population exactly in all respects. That is, there is uncertainty as to how well the sample results reflect the population. Statistical methods have been developed to reduce and quantify this uncertainty.

For example, an investigation into the performance of a new drug designed to alleviate the symptoms of a particular disease would involve taking a group of people suffering from this disease and inferring that the results of this trial would apply to all those suffering, and those who suffer in the future, from the disease.

Two aspects of statistical inference are estimation and hypothesis testing using statistical tests.

ESTIMATION

To obtain an accurate estimate of a population parameter, the sample must be representative of the population. To avoid bias, the sample items should be selected from the population at random. This means that all members of the population have an equal chance of being in the sample.

The precision of the estimate depends on the size of the sample. Clearly the larger the sample the better the estimate will be. Precision is measured by calculating the standard error of the estimate or a confidence interval (usually the 95% confidence interval).

Worked example   Consider the following times (to the nearest hour) that 16 patients experience relief from a migraine after taking a certain drug:

7	8	1	2	6	3	5	2	4	9	4	6	5	6	9	8

Mean = 5.312 hours

Standard deviation = 2.522 hours

Thus the estimate of the mean time for a patient to experience relief is 5.312 hours. The 95% confidence interval for the mean time to experience relief is calculated to be 3.975 to 6.649.

It can be said that there is a probability of 0.95 that the population mean lies between 3.975 hours and 6.649 hours. This provides a clear idea of how precisely the population mean has been estimated by these data.

The precision of estimates should always be reported alongside the estimate, and sometimes authors quote the standard error rather than a confidence interval.

Although 95% confidence intervals are most often reported, you will sometimes see 99% confidence intervals, in which case the confidence interval contains the population parameter with probability 0.99 and will, consequently, be wider than the corresponding 95% confidence interval. To calculate a 99% confidence interval, the factor 2 is replaced by 2.6.

Calculation of a 95% Confidence Interval for the Mean

The 95% confidence interval for a mean is calculated (approximately) from the following formula.

95% Confidence level for mean =

$$(\text{Sample mean} - 2) \times (\text{SE of mean to sample mean}) + (2 \times \text{SE of the mean})$$

where,

$$\text{Standard error of the mean} = \frac{\text{Sample standard deviation}}{\sqrt{\text{Sample size}}}$$

The factor 2 varies according to the sample size but only varies from 2.201 to 1.960 for sample sizes greater than 10, so that 2 is an adequate approximation in most cases. If you use a computer package that calculates the confidence interval the exact factor will be used.

Note: Since the sample mean and standard deviation are estimates of fixed (unknown) quantities, the only way of affecting the confidence interval is by altering the sample size, n. Increasing n will reduce the standard error of the mean and thus the width of the interval. But notice that to halve the width of the interval we have to quadruple the sample size (because of the square root in the formula).

PART II

ESSENTIALS OF WRITING

8

PLANNING THE THESIS

LOOKING FOR A RESEARCH TOPIC

The first step in understanding to do research is to identify the general areas in which you want to work. How do you find a suitable research topic? Ideas come from many diverse sources. These may include simple observations of everyday events; they may be the result of reading books and journals and asking "why?".

The four most common sources for finding a topic or problem include:

- one's experiences
- a review of previous research (literature review)
- a review of theories
- ideas from others

Performing a literature review is also a good method of developing a question. Picking a topic of interest and examining what research has been performed previously may be an effective way to do this. Reviewing published literature may provoke an idea or question by suggesting what areas of a topic should be explored further. Questions may also be raised by the inconsistencies that are suggested or by the lack of what exists about a topic in the published literature. In practice, the topic may be pre-determined by the choice of a supervisor. The choice of the supervisor has made or marred many a student's research career. Selecting the right supervisor is perhaps more important than working on a particular problem. You should be introduced to research by a stimulating and an understanding advisor. The best way to check out a potential advisor is to talk to her/his ex-students. The credentials of potential supervisors as researchers can be verified by going through their research publications.

Having identified a topic within your general field of interest, you must now discover the current state of knowledge in that area. It would indeed be very frustrating to find out towards the end that the question you are attempting to answer has been answered years ago. This can be avoided by conducting a thorough search of all relevant information available on the topic.

While finalizing the research topic, there are some points you must keep in mind:

1. The topic must interest both you and to some extent your supervisor.
2. The topic should not be too expensive.
3. The topic should be one that is certain to produce results within the given time.
4. Select a topic which will result in a published paper. publication of a good research paper is the best way to build a favourable professional reputation.

SELECTING A TOPIC

The selection of a suitable topic for a thesis or dissertation is in many ways the most difficult task. A thorough knowledge of a particular subject area is needed. Hence, at the postgraduate level, students are required to have completed a major in the subject before embarking on the thesis. At the master's and doctoral levels, it is necessary not only to have completed one's earlier course work, but also to have successfully completed an honours thesis and sometimes a preliminary or qualifying examination as well.

When developing the question that is going to be researched, it is necessary to ensure that the question being asked has significance, is researchable, has feasibility and is of interest to the researcher. It is important that the research being performed has an outcome that is meaningful.

Some questions that should be considered are:

- Is the problem an important one?
- Will the results change current practice?
- Will the results lead to better applications, procedures or outcomes?
- Will the findings broaden knowledge or understanding?
- Will the results have relevance to the topic?

Asking these questions may help establish the usefulness of the possible research question being asked.

The more one knows about a particular field, the more able one is to detect gaps in it and to recognize problem areas that require investigation. It is the ability to detect problems that the postgraduate student must develop, for every thesis or dissertation should set out to shed light on the solution to a particular problem.

One of the best sources of problems for investigation is at the cutting edge where research is being carried out. The closest one can get to this cutting edge is through direct contact with the personnel at a research institution. The research fellow, lecturer or professor who is active in research is usually a fund of research problems. The more problems investigated, the more problems emerge for investigation.

Criteria for Selecting a Topic

Once you have narrowed down your field of interest, and have identified several problem areas for possible investigation, there are a number of questions you should ask about the topic.

1. **Does the topic really interest you?** At the least, you will be engaged on your research project for 2–3 years and its successful completion will depend in part on your sustained interest. Lose this, and the task becomes the worst kind of chore.

2. **Can the topic be completed in the required time?** Some topics by their very nature require a period of time to elapse before data can be collected. Longitudinal growth studies and long-term attitude studies fall into this category.

3. **Is the necessary equipment available?** Specialized and expensive equipment is required for many studies. Unless you

have a reasonable assurance that necessary equipment will be available when you require it, it would be wise to consider another topic.

4. **Are library facilities sufficient?** Library facilities are essential for literary or analytical research studies. A particular topic may prove unsuitable simply because there is not ready access to the requisite source materials.

5. **Is the study feasible?** The questions of the availability of equipment, subjects, library facilities and time have been noted. However, another question to consider is whether the research techniques for testing a particular problem have been developed or are sufficiently refined. This question implies that the student should determine the techniques he/she intends using before embarking upon a study.

6. **Is the problem a significant one?** The time, effort and expense for tackling a particular problem are to be justified, i.e., how far will results from the research help to improve the existing knowledge and situation.

7. **Is competent guidance available?** This is another important factor one should consider. Guides or supervisors should be experts in the area of research they are guiding.

DESIGNING THE STUDY

There are certain elements of research design common to empirical and analytical study.

1. **Statement of hypothesis** In the empirical study, the problem is re-expressed in terms of a specific hypothesis to be tested. These should be clearly stated and the relationship to previous research is made clear.

2. **Statement of assumption** In every study, it is necessary to make assumptions. Wherever these are made, they should be stated clearly. Nothing should be taken for granted.

3. **Statement of the limitations of the study** The time allowed for a study and restriction of length in reporting it impose necessary limitations on each study. The limits of the proposed investigation should be clearly defined.

4. **Definition of terms** All terms should be defined. The interpretation of the findings of a study depends in part on the way terms were originally defined.

5. **Appropriateness of research design** In the empirical study, the statistical methods for testing a hypothesis need to be described and examined for appropriateness.

6. **Description of population and samples** Most studies require a random or representative sample to be drawn from some population. Questions to consider are whether the population and sample are adequately described and whether the method of sampling is appropriate.

7. **The control of error** The control of error is applied principally to experimental studies. The experimenter has to consider what variables are operating in a given situation. In the laboratory, it is usually possible to control all variables or sources of error.

8. **Reliability and validity** It is necessary to establish the variability and validity of test instruments. That is, whether the test performed provides consistent measurements and whether tests really measures what is claimed for it.

EVALUATION OF PREVIOUS RESEARCH

After the selection and definition of a problem for the purpose of research, the research workers should find out what aspects of the problem selected by them have already been investigated into, what techniques were employed by different research worksers and what progress has been made in the direction of the desired solutions. If they select an absolutely new problem which has never been tackled before, their research may become a pioneer study. They should try to seek direction and impetus by critically evaluating every research report dealing directly or indirectly with the chosen problem. They should also try to know in detail about all related research projects in progress, but not yet completed or reported. These works are likely to furnish them with valuable suggestions on good procedures and tried techniques.

Working Bibliography

This should include all items or research information available from past experience in field research, courses of study in the University, documents and reports, various books and research monographs, all related to the problem under study. These sources will not only furnish actual facts previously determined, which will constitute the basis of further research, it will also give valuable hints on the methods of interpreting findings, classifying and presenting facts and drawing conclusions. The bibliography should be relatively short as well as strictly to the point.

Research Reading

The importance of adequate and skilled reading for carrying out systematic research is now widely recognized, not only in locating the research problem, but also in analysing the procedures and collecting evidence, and thus critical reading of professional literature becomes indispensable. Knowledge of the thinking experience of other research workers has gained importance often through reading other works. Take notes as you read the literature. You are reading to find out how each piece of writing approaches the subject of your research, what it has to say about it, and how it relates to your own thesis. Usually, you won't have to read the whole text from first to last page. Learn to use efficient scanning -and -skimming reading techniques. The purpose of research reading is discovery—the discovery of true values. Here are some practical hints which may be useful for the beginners.

1. Read only relevant literature. Read the entire work or only those chapters, sections or paragraphs, which are pertinent to the scope of the inquiry.

2. Read original works and reports from authoritative sources.

3. Read with comprehension. Reading for research involves learning and one should try to understand the author's mind in the correct perspective.

4. Read in time. Read the selective material at such a speed that the entire work of reading is covered within a specified time at the researcher's disposal.

5. Read with regular rhythm. Do not force yourself to read more than what you can assimilate in one sitting.

Research Diary

An important source of raw material for research is the personal record of the research workers themselves. The personal experience of the research worker is usually recorded in a diary in the form of brief statements with respect to research methodology, results, etc.

9

USING THE LIBRARY

A library is a storehouse of information available to those who know how to use it. The only way you can learn to use a library is by using it, there is no other way. There are different types of libraries.

1. College or department library—provides a collection of advanced textbooks on the subject of your interest.

2. University library—generally caters to most needs of the researchers, through its collection of reference materials, journals and books.

3. Government-run national libraries—e.g. National Medical Library, National Science Library.

4. American and British libraries—four metropolitan cities in India, viz., Delhi, Bombay, Calcutta and Chennai have branches of these libraries. The British Council also has libraries in cities like Ahmedabad, Bangalore, Bhopal, Hyderabad, Lucknow, Patna, Pune, Ranchi and Trivandrum. These libraries provide access to computer-assisted literature search. They also help in obtaining copies of journals or articles that may not be available locally for a fee.

GROUND PLAN OF A LIBRARY

◎ General reading room.

◎ Stack room—books on different subjects are arranged subjectwise.

◎ Reference section—can be referred within the library, not available for issue.

◎ Journal section—periodicals and journals are kept.

◎ Catalogue section—records the details about every book or journal received by the library.

There are three types of catalogues in use in library:

- ◎ Author catalogue
- ◎ Title catalogue
- ◎ Subject catalogue

All contain the same information. They differ only in the criteria they use to categorize. The author catalogue uses the surname of authors as its basis. Subject catalogue lists holdings according to the subject matter of the books. The cards in the classified catalogue are arranged according to a code used by the library. This classified code consists of a combination of letters and numbers and this identifies the location of every book, journal or magazine within the library.

A library classification is a system of coding and organizing library materials (books, serials, audio-visual materials, computer files, maps, manuscripts, realia) according to their subject and allocating a call number to that information resource. The system of arrangement adopted by a library is to enable patrons to locate its materials quickly and easily. Classification systems in libraries generally play two roles. Firstly they facilitate subject access by allowing the user to find out what works or documents the library has on a certain subject. Secondly, they provide a known location for the information source to be located (e.g. where it is shelved).

Classifications may be natural (e.g. by subject), artificial (e.g. by alphabet, form, or numerical order), or accidental (e.g. chronological or geographic). They also vary in degree; some have minute subdivisions while others are broader. Widely used systems include the Dewey Decimal Classification, the Library of Congress Classification, the Bliss Classification, and the Colon Classification; special libraries may devise their own unique systems.

PURPOSE AND FUNCTION OF LIBRARY CLASSIFICATION

- ◎ It aids in storage of information in a helpful sequence.
- ◎ Retrieval of stored information on demand is possible.
- ◎ It helps to compile subject bibliography.

◎ It assures speedy location and restoration of books in the shelves.

◎ The classified arrangement displays several books written by different authors in each subject, thereby enabling the readers to select their books.

◎ In stock-taking, it enables a thorough, efficient and speedy verification of the library collection.

TYPES OF LIBRARY CLASSIFICATION

1. Decimal classification—The most widely used classification scheme by libraries throughout the world is the Dewey decimal classification

2. Subject classification of James Brown

3. Library of Congress classification

4. Colon classification of S.R.Ranganathan

5. Bibliographic classification of H.E. Bliss

DEWEY'S DECIMAL CLASSIFICATION

The system for organizing the contents of a library is based on the division of all knowledge into 10 groups. Each group is assigned 100 numbers. Subdivisions eventually extend into decimal numbers; for example, the history of England is placed at 942, the history of the Stuart period at 942.06, and the history of the English Commonwealth at 942.063. The system was first formulated in 1873 by Melvil Dewey. The DDC attempts to organize all knowledge into ten main classes. The ten main classes are further subdivided. Each main class has ten divisions and each division has ten sections. Hence the system can be summarized in 10 main classes, 100 divisions and 1000 sections. This system employs a numerical notation and divides the whole field of knowledge into two ten main classes.

000	Computer science, Information, and General works (encyclopaedias)
100	Philosophy and Psychology
200	Religion, Mythology
300	Social Sciences, Education
400	Language, Linguistics
500	Science, Life, Physical
600	Technology, Applied Sciences, Medicine, Business

LIBRARY OF CONGRESS CLASSIFICATION

The **Library of Congress Classification** (**LCC**) is a system of library classification developed by the Library of Congress. It is used by most research and academic libraries in the US. and several other countries. It consists of separate, mutually exclusive, special classifications, often having no connection save the accidental one of alphabetical notation. The arrangement roughly follows groupings of social sciences, humanities, and natural and physical sciences. It divides the field of knowledge into 20 large classes and an additional class for general works. Each main class has a synopsis that also serves as a guide. The resulting order is from the general to the specific and from the theoretical to the practical.

A	General Works
B	Philosophy, Psychology, Religion
C	Auxiliary Sciences of History
D	History: General and Old World
E	History: America
F	History: America
G	Geography, Anthropology, Recreation
H	Social Sciences
J	Political Science
K	Law
L	Education
M	Music and Books on Music
N	Fine Arts
P	Language and Literature
Q	Science
R	Medicine
S	Agriculture
T	Technology
U	Military Science
V	Naval Science
Z	Bibliography, Library Science, Information Resources (General)

Letter classes I, O, W, X and Y are not in standard use.

COLON CLASSIFICATION

Colon classification (CC) is a system of library classification developed by S.R. Ranganathan. It was the first ever faceted (or analytico-synthetic) classification. It is especially used in libraries in India. Its name "Colon classification" comes from the use of colons to separate facets in class numbers. However, many other classification schemes, some of which are completely unrelated, also use colons and other punctuations in various functions. They should not be confused with Colon classification.

This system of classification was brought out in 1933 by S.R. Ranganathan. This system is being followed in Benaras Hindu University, Universities of Delhi and Madras. This system uses all the 26 letters of the alphabets, each representing a discipline or disciplines.

A	Natural Sciences	M	Useful arts
B	Mathematics	N	Fine arts
BZ	Physical Sciences	O	Literature
C	Physics	P	Linguistics
D	Engineering	Q	Religion
E	Chemistry	R	Philosophy
F	Technology	S	Psychology, Social Sciences
G	Biology	T	Education
H	Geology	U	Geography
I	Botany	V	History
J	Agriculture	W	Political Sciences
K	Zoology	X	Economics
KZ	Animal husbandry	Y	Sociology
L	Medicine	Z	Law

The Colon classsification uses 42 main classes that are combined with other letters, numbers and marks in a manner resembling the Library of Congress Classification to sort a publication.

Facets

CC uses five primary categories, or facets, to further specify the sorting of a publication collectively called "PMEST."

, [[personality]]

; [[matter]] or property

: [[energy]]

. [[space]]

` [[time]]

Classes

The following are the main classes of CC, with some subclasses, the main method used to sort the subclass using the PMEST scheme and examples showing application of PMEST.

A   Generalia

1   Universe of Knowledge

2   Library Science

3   Book Science

4   Journalism

B   Mathematics

B1   Arithmetic

B13   Theory of Numbers

B2   Algebra

B23   Algebraic Equations

B25   Higher Algebra

B3   Analysis

B33   Differential Equations *[equation]* , *[degree]* , *[order]* : *[problem]*

B331,1,2:1 Numerical solutions (*:1*) of ordered (*33 1*) linear (*, 1*) second order (*,2*) differential equations

B37   Real Variables

B38   Complex Variables

B4   Other Methods

B6   Geometry

B7   Mechanics

B8   Physico-Mathematics

B9   Astronomical Mathematics

C   Physics

C1   Fundamentals of Physics

C2   Properties of Matter

C3   Sound

C4   Heat

C6   Electricity

C7   Magnetism

C8   Cosmic Hypothesis

D   Engineering

E   Chemistry

facets

    :1   General Chemistry

    :2   Physical Chemistry

    :3   Analytical Chemistry

    :33   Qualitative Chemistry

    :34   Quantitative Chemistry

    :35   Volumetric Chemistry

    :4   Synthetic Chemistry

    :5   Extraction Chemistry

E1   Inorganic Chemistry

E10   Group 0

E11   Group 1

E110   Hydrogen

E1109 Lithium

E111   Sodium

E2   Hydroxyl (Base)

E3   Acid

E4   Salt

E5   Organic Substance

F   Technology

G   Biology

H   Geology

H1   Mineralogy

H2	Petrology
H3	Structural Geology
H4	Dynamic Geology
H5	Stratigraphy
H6	Palaeontology
H7	Economic Geology
H8	Cosmic Hypothesis
HX	Mining
I	Botany
J	Agriculture

facets

:1	Soil
:3	Propagation
:4	Disease
:5	Development
:6	Breeding
:7	Harvest
:91	Nomenclature
:92	Morphology
:93	Physiology
:95	Ecology

material facets

,2	Bulb
,3	Root
,4	Stem
,5	Leaf
,6	Flower
,7	Fruit
,8	Seed
,97	Whole Plant
J1	Horticulture
J2	Feed
J3	Food

J4	Stimulant
J5	Oil
J6	Drug
J7	Fabric
J8	Dye
K	Zoology *(same facet schedule as I Botany)*
KZ	Animal Husbandry *(same facet schedule as I Botany)*
L	Medicine *[organ]:[problem],[cause]:[handling]*
Z3	Pharmacology *[substance].[action],[organ]*
LZ5	Pharmacopoeia
M	Useful arts
M7	Textiles *[material]:[work]*
Ð	Spiritual experience and Mysticism *[religion], [entity]: [problem]*
N	Fine arts
NA	Architecture *[style], [utility], [part] : [technique]*
ND	Sculpture *[style], [figure] ; [material] : [technique]*
NN	Engraving
NQ	Painting *[style], [figure] ; [material] : [technique]*
NR	Music *[style], [music] ; [instrument] : [technique]*
O	Literature
P	Linguistics
Q	Religion
R	Philosophy
S	Psychology
T	Education
U	Geography
V	History
W	Political
X	Economics
Y	Sociology
Z	Law

Example   The most commonly cited example of the Colon Classification is the classification for:

"Research in the cure of the tuberculosis of lungs by X-ray conducted in India in 1950s":

- Main classification is Medicine
  - (Medicine, L)
- Within Medicine, the Lungs are the main concern
  - (Medicine,Lungs, L.45)
- The property of the Lungs is that they are afflicted with Tuberculosis
  - (Medicine,Lungs;Tuberculosis, L,45;421)
- Tuberculosis is being performed (:) on, that is the intent is to cure (Treatment)
  - (Medicine,Lungs;Tuberculosis:Treatment, L,45;421:6)
- The matter that we are treating Tuberculosis with are X-rays
  - (Medicine,Lungs;Tuberculosis:Treatment;X-ray, L,45;421:6;253)
- And this discussion of treatment is regarding the Research phase
  - (Medicine,Lungs;Tuberculosis:Treatment;X-ray:Research, L,45;421:6;253:f)
- This Research is performed within a geographical space (.) namely India
  - (Medicine,Lungs;Tuberculosis:Treatment; X-ray:Research.India, L,45;421:6;253:f.44)
- During the time (´) of 1950
  - (Medicine,Lungs;Tuberculosis:Treatment; X-ray:Research.India'1950, L,45;421:6;253:f.44'N5)

BLISS BIBLIOGRAPHIC CLASSIFICATION

The **Bliss bibliographic classification** (or **BC** for short) is a library classification system that was created by Hendry E. Bliss (1870–1955), published in four volumes between 1940 and 1953. Although originally devised in the United States, it was more commonly adopted by British libraries than by American ones. Bliss wanted a classification system that would provide distinct rules, yet still be adaptable to whatever kind of collection a library might have, since different libraries have different needs. His solution was the concept of "alternative location", in which a particular subject matter could be put in more than one place, as long as the library made a specific choice and used it consistently.

Examples

GER   Biochemistry of muscles in animals

MN&,S   History of Finland in the 19th century

PWWbca,L   The building and equipment of the YWCA in Albany, New York

Bliss deliberately avoided the use of the decimal point because of his hatred of Dewey's system. Instead he used upper- and lower-case letters, numbers, and every typographical symbol available on his extensive and somewhat eccentric typewriter (which included both forward and reverse italics).

Author Card

```
001.42[15]Gurumani[1], N (1943[2]-    )

GUR[16]         Research Methodology[3]: for

        biological sciences[4]/by N. Gurumani[5].-

41717[17] Chennai[6]: M.J.P Publishers[7],2006[8].

                782p[9]. :ill[10]. ; 24[11]cm

                ISBN 81-8094-016-0[12]

                1. Research Methodology[13]I. Title[14]
```

1. Choice of access by the name of the author

2. Year of birth of the author

3. Title

4. Other title information

5. Statement of responsibility

6. Place of publication

7. Name of the publisher

8. Year of publication

9. Physical description, textual pages

10. Illustrations are provided

11. Size of the book

12. ISBN

13. Subject Heading (Tracing)

14. Added Entry for Title (Tracing)

15. Address - Class No.

16. Address: Book No.

17. Address: Accession No.

Title Card

Research and Writing[1]

378.242[19] Ramadass, P[2] (1947[3]–   )

RAM[20]   Research and Writing[4]: across

the disciplines[5]/ by P. Ramadass[6] and A. Wilson

61717[21] Aruni[7].- Chennai[8]:MJP Publishers[9], 2009[10].

xx[11], 250p[12]. ; 23[13] cm.

Includes glossary, references and index[14].

ISBN 978-81-8094-068-2[15]

1. Dissertations, Academic-Writing[16] I Aruni,

A. Wilson[17] II Title[18]

1. AE under the title of the book

2. Choice of access under the author

3. Year of Birth of the author

4. Title of the book

5. Other title information

6. Statement of responsibility to first author

7. Statement of responsibility to second mentioned author

8. Place of publication

9. Name of the publisher

10. Year of publication

11. Physical description-preliminary pages

12. Physical description-textual pages

13. Size of the book

14. Notes for glossary, etc.

15. ISBN

16. Subject Heading-1

17. Added Entry - second mentioned author

18. Added Entry for Title

19. Address - Class No.

20. Address: Book No.

21. Address: Accession No.

10

SCIENTIFIC WRITING

The key to scientific writing is clarity. Clarity should be a characteristic of any type of communication. When something is being said for the first time, clarity is essential. Scientific communication is a two-way process. A published scientific paper is useless unless it is neither received nor understood by its intended audience. Scientific writing is the transmission of a clear signal to a recipient. The words of the signal should be as clear, simple and well-ordered as possible. Scientific writing neither has the room nor the need for ornamentation. The second principal ingredient of a scientific paper should be appropriate language. The scientific knowledge must be communicated effectively, clearly in words of certain meaning. "The best English is that which gives the sense in the fewest short words."

WRITING AT THE TERTIARY LEVEL

There is an increasing tendency in formal education to place more emphasis on the submission of written work as part of other course requirements and for purposes of student assessment. At many colleges and universities, considerable weight is now being given to student progress throughout the year which is measured by regular tests and assignments, rather than an end-of-the-year, do-or-die examination.

ASSIGNMENTS AND TERM PAPER

Typically, written assignments and term papers are geared to course work covered by a series of lectures or tutorials. The student is assigned to write about a particular topic or a list of topics from which to choose. The topics set for written assignment should be a useful guide covering important contents of the course of study. Assignments encourage the students 1) to read critically in a particular content area, 2) to search and select from available materials, 3) to marshal their thoughts of a topic and 4) to submit to the discipline of communicating their thoughts about a topic, 5) to present the evidence which they have sifted and evaluated and 6) to arrive at certain conclusions.

PLANNING THE ASSIGNMENT

Planning an assignment involves the following criteria.

- Defining the problem
- Limiting the problem
- Time schedule
- Consulting source material
- Preparing a working bibliography
- Taking notes
- Writing the outline
- Writing the first draft

THESIS AND DISSERTATIONS

Students enrolled for an honours degree, postgraduate diploma or a higher degree are almost invariably required to submit a thesis or dissertation. There is a lack of precision and common agreement in the term thesis and dissertation and often are used interchangeably. A thesis is much more than a large term paper. It normally represents the culmination of a substantial piece of original work over a period of at least one year. The thesis is expected to make an original contribution to knowledge. Students writing theses are much more responsible for the selection and delimitation of their area of study than those writing an assignment or essay.

CONVENTIONS OF THESIS WRITING

An assignment is more limited in scope and shorter than a thesis and it is less likely to involve original research. Both assignments and theses require the use of scholarly style, and recommendations in this section apply equally to assignment and thesis writing. The format of thesis varies from institution to institution. The content of the thesis is important; the presentation of the assignment in a standard form is a discipline which is also vital to the acceptability of the thesis. Good research may be marred by poor reporting; proper presentation is an integral part of the whole project. Research can become a contribution to a field of knowledge only when it is adequately communicated. For this, precise writing is essential. A careful choice of words will serve to convey the exact meaning. The best word to express an idea is not necessarily the longest word.

Colloquial, conventional or other modes of expression found in a latest best-selling novel are inappropriate in a thesis. Scientific writing is not of a personal or conversational nature and for this reason, a third person is usually used. As a general rule, personal pronouns such as I, we, you, me, my, ours or us should not appear except in quotations. A thesis should not consist of the reporting of personal experience or opinion but should be a critical analysis of a problem of the presentation of evidence relating to that problem.

The thesis writer should aim at a higher level of readability. Sentences should not be too involved and complex. Scholarly writing is not crammed as jargon. Sweeping statements and exaggerated claims should be avoided. Statements must be suitably qualified. Sound reasoning and intellectual honesty are hallmarks of scholarly style. Quotations must be accurately cited and suitably acknowledged. The contribution of other writers must be duly acknowledged.

Since thesis recounts what is already been done, it should be written in the past tense. Accurate spelling is essential for scholarly writing. An authoritative dictionary should be consulted for correct spelling. Particular attention should be paid to grammar and punctuation. On no account should abbreviations such as '&' be used in a thesis (although abbreviations are acceptable in tables). Where there is more than one way to spell a word, the writer should aim at consistency, e.g. the use of 's' or 'z' in organize or organise.

SCIENTIFIC PAPER

A scientific paper is a written and published report describing original research results. The Council of Biology Editors (CBE), an authoritative professional organization (in biology, at least) dealing with such problems, defined primary publication as follows:

An acceptable primary scientific publication must be the first disclosure containing sufficient information to enable peers to assess observations, to repeat experiments and to evaluate intellectual processes. In other words, primary publication is the first publication of original research results, in a form whereby peers of the author can repeat the experiments and test the conclusions and in a journal or other source document readily available within a scientific community.

The scientific paper should have clear and distinctive component parts, viz., Introduction, Methods, Results and Discussion (hence, the acronym, IMRAD). Actually, the heading "Materials & Methods" may be more common than the simpler "Methods". The Introduction should start with a succinct statement of the subject and explain the purpose. Materials and Methods should give sufficient details so that another worker can repeat the procedure. The data in results section should be unified and should be in coherent sequence. Only such tables, photographs, drawings or charts that are necessary to clarify and document the text should be used. Extensive discussion should be avoided. The discussion section should relate the new findings to previous results and include logical deductions. The abstract should summarize only the major results and conclusions.

11

THESIS OR ASSIGNMENT WRITING

By following stringent format requirements, the writers can not only systematize and structure their own thinking in terms of theme, unity and clarity, but can also facilitate the reading and interpretation of their work by others.

Generally, the mechanical format of a paper consists of three parts: the preliminaries, the text and the reference materials. The length of any of these three parts is conditional on the extent of their study. The order in which individual items within the three main sections appear is outlined below. This order should be strictly followed, although not every paper includes all the items listed.

The Preliminaries

1. Title page
2. Abstract
3. Preface, including Acknowledgement
4. Table of Contents
5. List of Tables
6. List of Figures or Illustrations

The Text

1. Introduction
2. Main body of the report (Usually divided into chapters and sections like Review of Literature, Materials and Methods, Results, Discussion, including Conclusions)
3. Summary

End matter

1. Bibliography
2. Appendix (or Appendices)
3. Index (if any)

PRELIMINARIES

Title Page

Most universities and colleges prescribe their own form of title page for theses, dissertations and research papers.

Written assignments will have the following in the title page.

1. Title of the assignment
2. Name of the writer
3. Name of the course for which the assignment was written
4. Name of the department
5. Name of the University, College and Institution
6. Date on which the paper is due

A thesis will have the following in the title page.

1. Title of thesis
2. Name and designation of the writer (optional)
3. Degree for which the thesis is presented
4. Name of the Institution/University to which the thesis is being submitted
5. Name of the department and institution
6. Month and year in which the thesis is submitted

Preface

The preface (often used synonymously with foreword) may include: the writer's purpose in conducting the study, a brief resume of the background, scope, purpose, general nature of the research upon which the report is being based and acknowledgments. For written assignments and minor dissertations, the preface may be omitted. If the writer has little of significance to say about his thesis that is not already covered in the main body of his report, the preface may also be omitted at the master's or doctoral level. In this eventuality, the page should be labelled "Acknowledgements" rather than "Preface".

EPIDEMIOLOGY OF RINDERPEST IN TAMIL NADU

By

K. Shankar

Research Methodology (RMC 601)

Department of Animal Biotechnology

Madras Veterinary College

Tamil Nadu Veterinary and Animal Sciences University

7th January, 1997

Figure 11.1   Sample title page for a written assignment

DEVELOPMENT OF A RECOMBINANT
VACCINE FOR RABIES

By

Dr. S. Kumar, M.V.Sc.

Thesis submitted in partial fulfillment of the requirement for

the degree of

Doctor of Philosophy

in

Animal Biotechnology

to the

Tamil Nadu Veterinary and Animal Sciences University

Chennai 600 007

Department of Animal Biotechnology

Madras Veterinary College

1997

Figure 11.2   Sample title page of a thesis

Acknowledgements recognize the persons to whom the writer is indebted for guidance and assistance during the study, and give credit to institutions for providing funds to implement the study or for use of personnel, facilities and other resources. For a term paper or written assignment, it is not necessary to acknowledge staff or institutions.

How to state the acknowledgements As to the Acknowledgements, two possible ingredients require consideration. First, you should acknowledge any significant technical help that you received from any individual, whether in your laboratory or elsewhere. You should also acknowledge the sources of special equipment, cultures, or other materials. You might, for example, say something like "Thanks are due to J. Jones for assistance with the experiments and to R. Smith for valuable discussion". Second, it is usually the Acknowledgements wherein you should acknowledge any outside financial assistance, such as grants, contracts or fellowships.

Advisor(s) and anyone who helped you technically (including materials, supplies), intellectually (assistance, advice) or financially (for example, departmental support, travel grants) should be acknowledged.

The important element in Acknowledgements is simple courtesy. There is not anything scientific about this section of a scientific paper. I wish that the word "wish" would disappear from Acknowledgements. Wish is a perfectly good word when you mean wish, as in "I wish you success". However, the word "wish" is an unacceptable substitute for the word "want". If you say "I wish to thank John Jones", you are wasting words. "I thank John Jones" is sufficient.

Table of Contents

The Table of Contents includes the major divisions of the thesis: the introduction, the chapters with their subsections and the bibliography and appendix. Page numbers for each of these divisions are given. Care should be exercised that titles of chapters and captions of subdivisions within chapters correspond exactly with those included in the body of the report. It is optional whether the Title page, Acknowledgements, List of Tables and List of Figures are included in the Table of Contents.

TABLE OF CONTENTS

ABSTRACT

ACKNOWLEDGEMENTS   ii

LIST OF TABLES   vi

LIST OF FIGURES   viii

Chapters

    1.   INTRODUCTION   1
    2.   REVIEW OF LITERATURE   4
    3.   METHODS AND MATERIALS   30
    3.1
    3.1.1
    3.2
    3.3
    3.4
    4.   RESULTS
    4.1
    4.2
    4.2.1
    4.3
    4.4
    5.   DISCUSSION
    5.1
    5.2
    5.2.1
    5.3
    5.4   CONCLUSIONS
    6.   SUMMARY
            BIBLIOGRAPHY
            APPENDICES

Figure 11.3   Sample table of contents for a thesis

List of Tables

After the Table of Contents, the List of Tables is given. The heading LIST OF TABLES, should be centred on a separate page by itself. Two spaces below this, the headings "Table" and "Page" appear at the left and right margins, respectively. For each table, the number of the table in Arabic/Roman numerals must appear, the exact caption or title of the table and the page number.

LIST OF TABLES

Table No.	Table Caption	Page
1.	The monthwise distribution of disease	67
2.	Seasonal incidence rate	68
3.	Particular of samples collected	69
4.	Comparison of various test results	70

Figure 11.4   A sample list of tables

List of Figures

The List of Figures appears in the same way as the List of Tables. The page is headed LIST OF FIGURES. Figures are listed at the left of the page under heading Figures.

The term Plates are often used for photographs.

LIST OF FIGURES

Figure No.	Figure Caption	Page
1.	The typical ELISA reaction	75
2.	Cytopathogenic effect in cell lin	78
3.	PCR amplification of N-gene	81
4.	Restriction profiling of H-gene	82
5.	Electropherogram of the sequenced data	85

Figure 11.5 A sample list of figures

THE TEXT

The thesis proper follows the preliminaries detailed above and begins with the first page of the text. The text is the most important part of a thesis as it is in this section that the facts are presented. The writers should devote a greater part of their time and energy to a careful organization and presentation of their findings or general arguments.

Introduction

An introduction should be written with considerable care with two major aims in view: introducing the problem in a suitable context and arousing and stimulating the reader's interest. If introductions are dull, aimless, confused, rambling and lacking in precision, direction and specificity, there is little incentive for the reader to continue reading. An introductory chapter usually contains the following:

i.   A lucid, complete and concise statement of the problem being investigated or the general purpose of the study.

ii. A justification for the study, establishing the importance of the problem.

iii. A preview of the organization of the rest of the paper or thesis to assist the reader in grasping the relationship between the various parts of the paper.

Main Body of the Report

There are certain general principles which should be followed:

1. Organize the presentation of the argument or findings in a logical and orderly way, developing the aims stated or implicated in the introduction.

2. Substantiate arguments or findings.

3. Be accurate in documentation.

In both theses and assignments, every effort should be made to write clearly and forcefully within a logical framework. This section includes Review of Literature, Materials and Methods, Results and Discussion.

Conclusions

The conclusions serve the important function of tying together the whole thesis or assignment. In summary form, the developments of the previous chapters should be succinctly restated, important findings discussed and conclusions drawn from the whole study. In addition, the writer may list "unanswered questions" that have occurred in the course of the study and which require further research beyond the limits of the project being reported. The conclusion should leave the reader with the impression of completeness and of positive gain.

END MATTER

Bibliography or Reference

This should be capitalized and centred. The bibliography follows the main body of the text and is a separate but integral part of a thesis. Pagination is continuous and follows the page numbers in the text. Complete details of all the references cited in the text should be listed, including scientific papers, monographs, proceedings of conferences, textbooks, etc. For assignments, the heading REFERENCES may be adequate.

Appendix

It is usual to include in an appendix such matters as original data tables that present supporting evidence, tests that have been constructed by the research student, parts of documents or any supportive evidence that would detract from the major line or argument and would make the body of the text unduly large and poorly structured.

The following can be included in the appendix:

◎ Reference data/materials not easily available (these are used as a resource by the department and other students).

◎ Tables (more than 1–2 pages long).

◎ Calculations (more than 1–2 pages long).

◎ A key article.

◎ If you consulted a large number of references but did not cite all of them, you might want to include a list of additional resource material, etc.

- List of equipment used for an experiment or details of complicated procedures.

Note   Figures and tables, including captions, should be embedded in the text and not in an appendix, unless they are more than 1–2 pages and are not critical to your argument.

Index

If an index is included, it follows the bibliography and the appendix. An index is not required for a written assignment or for an unpublished thesis. If a thesis is subsequently published as a book, monograph or bulletin, an index is necessary for any work of complexity.

12

PREPARING THE TITLE

In preparing a title for a paper or thesis, the author would do well to remember one salient factor: that the title will be read by thousands of people. Perhaps, few people will read the entire paper, but many people will read the title, either in the original journal or in one of the secondary (abstracting and indexing) services. Therefore, all words in the title should be chosen with great care, and their association with one another must be carefully managed.

What is a good title? It is defined as the "fewest possible words that adequately describe the contents of the paper". The title should not be too lengthy. It should not be too short also. Titles like "Studies on *Brucella*" are not very helpful to the potential reader. Was the study taxonomic, genetic, biochemical or medical? Most of the time, long titles contain "waste" words. Often, these waste words appear right at the start of the title. Words such as "Studies on", "Investigations on" and "Observations on" should be avoided.

Let us analyse a sample title: "Action of antibiotics on bacteria". Is it a good title? In form it is. It is short and carries no excess waste words. We can safely assume that the study introduced by the above title did not test the effect of all antibiotics on all kinds of bacteria. Therefore, the title is essentially meaningless. If only one or few of the antibiotics or microorganisms were studied, they should be individually listed in the title. Examples of more acceptable titles are:

"Action of streptomycin on *Mycobacterium tuberculosis*"

"Action of streptomycin, neomycin and tetracycline on gram-positive bacteria"

"Action of various anti-fungal antibiotics on *Candida albicans* and *Aspergillus fumigatus*"

Although these titles are more acceptable than the sample, they are not especially good because they are still too general. If the "action of" can be defined easily, the meaning might be clearer. For example, the first title above might be rephrased as "Inhibition of growth of *Mycobacterium tuberculosis* by streptomycin".

The title of a paper is a label. It is not a sentence. Because it is not a sentence, it is really simpler than a sentence (or usually shorter). The meaning and order of the words in the title are of importance to the potential reader who sees the title in the journal or table of contents.

As an aid to the reader, "running titles" or "running heads" are printed at the top of each page. Often the title of the journal or book is given at the top of left-facing pages and the article or chapter title is given at the top of right-facing pages. The titles should almost never contain abbreviations, chemical formulae, proprietary names and jargon.

13

WRITING ABSTRACT AND SUMMARY

An abstract should be viewed as a mini-version of the paper. The abstract should provide a brief summary of each of the main sections of the paper: Introduction, Materials and Methods, Results and Discussion. Thesis abstract should focus more on problem and solution discussed in the thesis. Methodology of collection of data is not needed, nor is it needed that the researcher mentions literature reviews. A well-prepared abstract enables readers to identify the basic content of a document quickly and accurately, to determine its relevance to their interests, and thus to decide whether they need to read the document in its entirety. The abstract should not exceed 250 words and should be designed to define clearly what is dealt within the paper. The abstract should

1. state the principal objectives and scope of the investigation,
2. describe the methodology employed,
3. summarize the results and
4. state the principle conclusions.

An Abstract should be

1. as short as possible,
2. should get straight to the heart of the article, ignoring the lengthy introduction/justification; give conclusions and findings, not simply an outline of the contents, aims or methods,
3. should be written in full sentences, not in note form and
4. should include keywords.

The key elements of Abstract are

B   Background information (present tense)
P   Principle activity (purpose) (past tense/present perfect tense)
M   Methodology (past tense)
R   Results (past tense)
C   Conclusions (present tense/tentative verbs)

Most or the entire abstract should be written in the past tense, because it refers to work done. Abstract should never give any information or conclusions that are not stated in the paper. Reference to the literature must not be cited in the Abstract (except in rare instances, such as modification of previously published methods). There are two types of Abstracts:

Informative abstract   This is usually used in Primary journals and is designed to capsulize the paper. It should briefly state the problem, the method used to study the problem and the principal data and conclusions.

Indicative abstract   This is also known as descriptive abstract. This type of abstract is designed to indicate the subject of a paper, making it easy for potential readers to decide whether to read the paper. However, because of its descriptive rather than substantive nature, it can seldom serve as a substitute for the full paper. Thus, indicative abstracts should not be used as "headings" in research papers.

Today most scientific journals print a heading abstract with each paper. It is printed as a single paragraph. Abstract should be written clearly and simply. If you cannot attract the interest of the reviewer in your abstract, your cause may be lost. When writing an abstract, examine every word carefully. If you can tell your story in 100 words, do not use 200. Economically, it does not make sense to waste words. It costs about Rs.4/- a word to publish a scientific paper and another Rs. 4/- every time that word is reprinted in an abstracting publication.

In a thesis abstract, the researcher should try to stuff the abstract with keywords. Almost all the research work is stored in electronic form now. The computers need keywords to search for appropriate work. The more keywords a researcher includes in the thesis abstract, the more people will be able to access the work. A good thesis abstract can also be used as the introduction or opening section for the actual thesis.

SUMMARY

Summary is different from abstract that appears at the beginning of the paper, and usually summarizes all major parts of a paper. Summary is placed at the end of the paper. The summary should satisfy the following conditions:

- It should summarize the report in such a way that it could stand on its own and would make sense to a managerial, public, or non-technical audience.

- It should include just enough background and information about methodology to orient readers.

- It should summarize the key information from the report, concentrating on the problem, conclusions, and recommendations. It should also include essential "bottom line" figures on cost or performance on which the recommendations are based.

- It should exclude any information not presented in the report itself.

Summary generally addresses a technical audience and concisely presents key information about the study, its methodology, its principal results, and their significance. Some reports may contain both an executive summary and a technical abstract.

14

WRITING THE INTRODUCTION

The first section of the text proper is the Introduction. The Introduction should introduce the problem in a suitable context and it should arouse and stimulate the reader's interest. If the Introduction is dull, aimless, confusing and lacking in precision, direction and specificity, there is little incentive for the reader to continue reading.

There is a saying: "A bad beginning makes a bad ending".

The Introduction should introduce the thesis. This is not a summary of the thesis. It is not a brief version of each chapter. It is an introduction to the topic. Introduce the subject. In general terms, what does your study address? Why is it important? Where does it fit in the overall field? Be sure to include in the Introduction a clear statement of your hypothesis and how you are going to address it. Be sure to include a hook at the beginning of the Introduction. This is a statement that is sufficiently interesting to motivate your reader to read the rest of the paper, it is an important/interesting scientific problem that your paper either solves or addresses. You should draw the reader in and make them want to read the rest of the paper.

The purpose of the Introduction should be to supply sufficient background information to allow the reader to understand and evaluate the results of the present study without needing to refer to previous publications on the topic. The introduction should also provide the rationale for the present study. Above all, you should state briefly and clearly your purpose in writing the paper. You should include a clear statement of your hypothesis and how you are going to address it. Reference should be chosen carefully to provide the most important background information. Much of the introduction should be written in the present tense, because you will be referring primarily to your problem and the established knowledge relating to it at the start of the work.

There are five main stages in the writing of the Introduction:

◎ **1st Stage**   General statement (s) about a field of research to provide the reader with a setting for the problems to be reported. You should present relevant literature that supports the need for your project. Research articles, books, educational and government statistics are just a few sources that should be used here.

◎ **2nd Stage**   More specific statements about the aspects of the problems already studied by other researchers. You should succinctly state the problem that you are going to address. You should also present relevant information about why this is an important problem.

◎ **3rd Stage**   Statement(s) that indicate the need for more investigations. You should carefully present the model or theory that underlies the project. The rationale should define the larger problem being investigated, summarize what is known about the problem, define the gap(s) in the knowledge, and state what needs to be done to address the gap(s).

◎ **4th Stage**   Very specific statement(s) giving the purpose/objectives of the writer's study. Based on the above background information, explain the purpose of the study. Explain what you hope the study will accomplish and why you chose to do this particular study. This should be supported with citations and specific information related to the study. You state the hypothesis(-ses) that will be tested in your study.

◎ **5th Stage**   Optional statement(s) that give value or justification for carrying out the study.

The purpose of the Introduction is to introduce the paper or thesis to the reader. Unless the problem is not stated in a reasonable, understandable way, readers will have no interest in your paper. The introduction should have a hook to gain the reader's attention. Why did you choose that subject and why it is important, should be clearly stated.

In the 3rd stage, existing gaps in knowledge are to be indicated.

1. You may indicate the need for research in particular area, where "very little work has been done" so far;

2. You may indicate that an examination of the previous literature raises new questions not previously considered by other workers;

3. You may indicate that there is an unresolved conflict among the authors of previous studies which is to be resolved.

Signal word	Gap	Research topic
	(present/present perfect)	
However/But	few studies have been done	
	little literature is available on	
	very little is known about	
	no work has been done on	

In the 4th stage, the statement of purpose is given

The purpose of this thesis The aim of the present paper The object of this report	is to determine	
This paper	describes presents	(present tense)
The thesis will	deal with discuss	(future tense)

The Introduction should include the following information.

1. A statement of the goal of the paper—why the study was undertaken, or why the paper was written. Do not repeat the abstract.

2. Sufficient background information to allow the reader to understand the context and significance of the question you are trying to address.

3. Proper acknowledgement of the previous work on which you are building. Sufficient references such that a reader could, by going to the library, achieve a sophisticated understanding of the context and significance of the question.

4. Explain the scope of your work, what will and will not be included.

The Introduction is where you "soft launch" your reader on the work described in your thesis. Lead the reader from the known to the unknown. State the hypothesis clearly. Give a preview of your thesis, globally and chapter by chapter. Your Introduction has done its work if you have captured the reader's curiosity and interest in this first chapter.

15

WRITING REVIEW OF LITERATURE

Review of literature gives an overview of the field of study. The literature review should begin with a reiteration of the purpose of your study. This should be followed by a preview of what is to come in the literature review. The purpose of the literature review is to concisely demonstrate your level of understanding of the research related to your project. You should not have an in-depth discussion of the literature. Rather you should group your literature according to some general topics and only discuss specific studies if they are "landmark" studies for your area of research. It should study the prevailing theories and hypotheses, and the key writers in the field.

A review of literature has the following functions:

- to justify your choice of research question, theoretical or conceptual framework, and method;
- to establish the importance of the topic;
- to provide background information needed to understand the study;
- to show readers you are familiar with significant and/or up-to-date research relevant to the topic;
- to establish your study as one link in a chain of research that is developing knowledge in your field.

The review traditionally provides a historical overview of the theory and the research literature, with a special emphasis on the literature specific to the thesis topic. It serves as well to support the argument/

proposition behind your thesis, using evidence drawn from authorities or experts in your research field.

A literature covers everything relevant that is written on a topic: books, journal articles, newspaper articles, historical records, government reports, theses and dissertations, etc. The important word is "relevant". In writing the literature review, your purpose is to convey to your reader what knowledge and ideas have been established on a topic, and what their strengths and weaknesses are. As a piece of writing, the literature review must be defined by a guiding concept (e.g. your research objective, the problem or issue you are discussing, or your argumentative thesis). It is not just a descriptive list of the material available, or a set of summaries. Besides enlarging your knowledge about the topic, writing a literature review lets you gain and demonstrate skills in two areas:

1. **information seeking**—the ability to scan the literature efficiently, using manual or computerized methods, to identify a set of useful articles and books and

2. **critical appraisal**—the ability to apply principles of analysis to identify unbiased and valid studies.

A thorough literature review is essential because it shows that you have studied rigorously what others have done. This lends credibility when you state the problem the dissertation is addressing, and when you provide reasons as to why obtaining a solution is important. The literature review should end with a discussion of how the literature relates to your study. Review of literature study is useful in the following ways:

- to have an overview of the field of inquiry
- to know what has already been said on the topic
- to know who the key writers are
- to know what the prevailing theories and hypotheses are
- to know what questions are being asked
- to know what methodologies and methods are appropriate and useful
- to avoid duplication of work
- to identify research problems
- to get complete background information on the topic pertaining to the field of research investigation

When writing the literature review

◉   Include only those works that are relevant to your research.

◉   Make sure that you have read and understood cited work.

◉   Organize your content according to ideas instead of individual publications.

Linking words are important. If you are grouping together writers with similar opinions, you would use words or phrases such as *similarly, in addition, also, again.*

More importantly, if there is disagreement, you need to indicate clearly that you are aware of this by the use of linkers such as *however, on the other hand, conversely, nevertheless.*

At the end of the review you should include a summary of what the literature implies, which again links to your hypothesis or main question.

SOURCE OF LITERATURE

1. ***Primary sources***   This gives the first-hand source of information of experimentation, e.g. original articles contained in research journals. For a research student, it may be very essential. Also, primary sources include reports, theses and conference proceedings.

2. ***Secondary sources***   These are summaries of information gathered from primary sources as in the case of *Veterinary Bulletin* which gives only abstracts or translations if it is in another language. Abstracts and review of articles are weak in giving information. Secondary sources give only fragmented information, and are not complete. It is of screening in nature. Other sources include abstracts, encyclopaedias, dictionaries, indexes and reviews.

3. ***Tertiary sources***   Information contained in the textbooks which are pooled information obtained from primary and secondary sources are the tertiary sources. The information is not very reliable because it may not be accurate as it changes when passed from one person to another. Directories and Guides are included in this.

Literature scanning is a very important aspect of research. Every year, about 3 million publications come out in the world. So, literature is vast. In India, 40 research publications are available in the Animal Husbandry area.

Indexes

Indexes of journals are the first place of scanning, i.e., they are periodicals with the list of articles arranged in alphabetical order under each subject. This gives information on sources of articles, titles and names of authors, e.g. Index Veterinarius, Index Medicus. Indexes are alphabetical listings of subjects and/or authors. Current Contents is a weekly publication that reproduces a photocopy of the contents page of current issues of most journals. No summaries are given, but this weekly publication lets you know in advance the papers of interest likely to be published in the forthcoming issue of journals.

Reviews

A review is an integrated and organized discussion of the literature pertaining to a well-defined subject. It usually covers a limited period of time. The reviewing of progress in specific areas of subjects are valuable at the start of a research project.

Cumulative Book Index

It gives titles of all books published in the world, subjectwise. It gives all books published from 1980 onwards. So, we can find out which are the new books. It is updated every year. It is published in the UK in English.

Abstracting Journals

This contains summaries of articles with authors and titles.

Examples are

- *Veterinary Bulletin*
- *Biological Abstracts*
- *Tropical Disease Bulletin*
- *Dissertation Abstracts*
- *International Abstracts of Biological Sciences*

Examples of specialized abstracting journals are

- *Thesis Abstracts*
- *Animal Breeding Abstracts*
- *Helminthology Abstracts*

- ◎ *Poultry Abstracts*
- ◎ *Virology Abstracts*
- ◎ *Bacteriological Abstracts*
- ◎ *Genetic Abstracts*
- ◎ *Mycology Abstracts*

Encyclopaedias

These give articles on a number of subjects written by specialists. Broad outlines of all subjects are given.

Encyclopaedia Brittanica	24 volumes
Encylcopaedia Americana	33 volumes

Year Book

This contains information on the research developments taking place during a year in a particular field, e.g. *Pathology Year Book; Pharmacology Year Book.*

Periodicals

These are publications issued in successive parts usually at regular intervals.

1. Weeklies—*Nature, Science, Veterinary Record, Lancet, New Scientist*
2. Monthly—*IVJ, Poultry Guide, Veterinary Microbiology*
3. Bimonthly—*J. of Extension, J. Hereditary, Veterinary Pathology*
4. Quarterly—*Meat Science, Microbiology Review, Avian Pathology, Comp. Immunol. Microbiol. Infect. Disease*
5. Half yearly—*Kerala J. Vet. Sci., J. Parasitology, J. Helminthology*
6. Yearly—*Advances in Vet. Science, Poultry Industry Year Book*

COMPUTER-AIDED SEARCHES

Database

This is an organized collection of information. There are a number of databases we are already familiar with. These include the telephone directory, dictionaries, encyclopaedia and card catalogues in the libraries. The term database used here means a computer program that is designed to store and retrieve information of all kinds. Data can be entered in any order and then sorted out using any factor stored.

Computer Searches

There are two kinds of computer searches: online and offline search. An online search is the one that conducts a direct dialogue between the searcher and the host. The questions are transmitted to the data bank from a computer terminal through the telephone line using a modem and the answers are sent back very quickly. This is a very expensive method because every minute of the total duration of contact has to be paid for.

Offline search uses a CD-ROM version of the database. A CD-ROM (Compact Disk Read Only Memory) is quite like a floppy disk, but with two main differences. Unlike a floppy disk, information on the CD-ROM cannot be altered. It is a permanent record. Further, the storage capacity of a CD-ROM cannot be altered. It is a permanent record. Further, the storage capacity of a CD-ROM is over 500 megabytes (=1200 diskettes). If the information from one CD-ROM were to be printed, it would fill 2,16,000 A4 size papers. Apart from the cost, the difference between an online and CD-ROM search is that the former is up-to-date.

Offline	BIOSIS (Biosciences)
	MEDLINE
	VETCD, Beast CD
Online	ERNET (Education and Resources)
	NICNET
	MEDLERS (PubMed)
	INTERNET
	DIALOG

SEARCH ENGINES

Search engine is an information retrieval system designed to help find information stored on a computer system, such as on the World Wide Web, inside a corporate or proprietary network, or in a personal computer. The search engine allows one to ask for content meeting specific criteria (typically those containing a given word or phrase) and retrieves a list of items that match those criteria. This list is often sorted with respect to some measure of relevance of the results. Search engines use regularly updated indexes to operate quickly and efficiently. Without further qualification, **search engine** usually refers to a **Web** search engine, which searches for information on the public Web. Other kinds of search engines are **enterprise search engines**, which search on intranets, personal search engines, and mobile search engines. Different selection and relevance criteria may apply in different environments, or for different uses. Some search engines also mine data available in newsgroups, databases, or open directories. Unlike Web directories, which are maintained by human editors, search engines operate algorithmically or are a mixture of algorithmic and human input.

Google

Around 2001, the Google search engine rose to prominence. Its success was based in part on the concept of link popularity and PageRank. The number of other websites and web pages that link to a given page is taken into consideration with PageRank, on the premise that good or desirable pages are linked to more than others. The PageRank of linking pages and the number of links on these pages contribute to the PageRank of the linked page. This makes it possible for Google to order its results by how many websites link to each found page. Google's minimalist user interface is very popular with users, and has since spawned a number of imitators. Google and most other web engines utilize not only PageRank but more than 150 criteria to determine relevance. The algorithm "remembers" where it has been and indexes the number of cross links and relates these into groupings. PageRank is based on citation analysis that was developed in the 1950s by Eugene Garfield at the University of Pennsylvania. Google's founders cite Garfield's work in their original paper. In this way virtual communities of web pages are found. Teoma's search technology uses a communities approach in its ranking algorithm. Nippon Electric Corporation (NEC) Research Institute has worked on similar technology. Web link analysis was first developed by Jon Kleinberg and his team while working on the CLEVER project at IBM's Almaden Research Center. Google is currently the most popular search engine.

Yahoo! Search

The two founders of Yahoo!, David Filo and Jerry Yang, Ph.D. candidates in Electrical Engineering at Stanford University, started their guide in a campus trailer in February 1994 as a way to keep track of their personal interests on the Internet. Before long they were spending more time on their home-brewed lists of favourite links than on their doctoral dissertations. Eventually, Jerry and David's lists became too long and unwieldy, and they broke them out into categories. When the categories became too full, they developed subcategories ... and the core concept behind Yahoo! was born. In 2002, Yahoo! acquired Inktomi and in 2003, Yahoo! acquired Overture, which owned AlltheWeb and AltaVista. Despite owning its own search engine, Yahoo! initially kept using Google to provide its users with search results on its main website Yahoo.com. However, in 2004, Yahoo! launched its own search engine based on the combined technologies of its acquisitions and providing a service that gave pre-eminence to the Web search engine over the directory.

Microsoft

The most recent major search engine is MSN Search (evolved into Live Search), owned by Microsoft, which previously relied on others for its search engine listings. In 2004 it debuted a beta version of its own results, powered by its own web crawler (called msnbot). In early 2005 it started showing its own results live. This was barely noticed by average users unaware of where results come from, but was a huge development for many webmasters, who seek inclusion in the major search engines. At the same time, Microsoft ceased using results from Inktomi, now owned by Yahoo!. In 2006, Microsoft migrated to a new search platform, Live Search, retiring the "MSN Search" name in the process.

HOW SEARCH ENGINES WORK

A search engine operates by the following three steps.

1. Web crawling
2. Indexing
3. Searching

Web search engines work by storing information about a large number of web pages, which they retrieve from the WWW itself. These pages are retrieved by a Web crawler (sometimes also known as a spider)—an automated Web browser which follows every link it sees. Exclusions can be made by the use of robots.txt. The contents of each page are then analysed to determine how it should be indexed (for example, words are extracted from the titles, headings, or special fields called meta tags). Data about web pages are stored in an index database for use in later queries. Some search engines, such as Google, store all or part of the source page (referred to as a cache) as well as information about the web pages, whereas others, such as AltaVista, store every word of every page they find. This cached page always holds the actual search text since it is the one that was actually indexed, so it can be very useful when the content of the current page has been updated and the search terms are no longer in it. This problem might be considered to be a mild form of linkrot and Google's handling of it increases usability by satisfying user expectations that the search terms will be on the returned webpage. This satisfies the principle of least astonishment since the user normally expects the search terms to be on the returned pages. Increased search relevance makes these cached pages very useful, even beyond the fact that they may contain data that may no longer be available elsewhere.

When a user enters a query into a search engine (typically by using keywords), the engine examines its index and provides a listing of best-matching web pages according to its criteria, usually with a short summary containing the document's title and sometimes parts of the text. Most search engines support the use of the boolean operators AND, OR and NOT to further specify the search query. Some search engines provide an advanced feature called proximity search which allows users to define the distance between keywords.

The usefulness of a search engine depends on the relevance of the result set it gives back. While there may be millions of web pages that include a particular word or phrase, some pages may be more relevant, popular, or authoritative than others. Most search engines employ methods to rank the results to provide the "best" results first. How a search engine decides which pages are the best matches, and what order the results should be shown in, varies widely from one engine to another. The methods also change over time as Internet usage changes and new techniques evolve. Most Web search engines are commercial ventures supported by advertising revenue and, as a result, some employ the controversial practice of allowing advertisers to pay money to have their listings ranked higher in search results. Those search engines which do not accept money for their search engine results make money by running search-related ads alongside the regular search engine results. The search engines make money every time someone clicks on one of these ads.

SOURCES OF INFORMATION FOR A LITERATURE REVIEW

There are a wide variety of resources available to assist you in locating articles for your review of related literature. Many of these are available through your college or public library in computerized database form.

1. *Education Index*—covers professional publications, precedes the use of CIJE (Current Index to Journals in Education).

2. *Reader's Guide to Periodical Literature*—covers articles from 200 widely read magazines (popular literature).

3. *Expanded Academic Index*—covers journal articles from over 1500 periodicals.

4. *Dissertation Abstracts International*—contains bibliographic citations and abstracts from doctoral disserations and master's theses worldwide.

5. *Psychological Abstracts*—presents summaries of studies completed in psychology, including developmental psychology and educational psychology. These two areas are of special interest to educational researchers.

6. ERIC (The Educational Resources Information Center)—collects and disseminates reports of current educational research, evaluation, developmental activity. ERIC maintains two databases searchable by computer. ERIC is probably the single most important source of information for educational researchers.

    i.   *RIE (Resources in Education)*—this ERIC database contains bibliographic citations and summaries to information not published in journals, e.g. conference presentations, technical reports, and unpublished research results. The items in this database are identified by ED numbers.

    ii. *CIJE (Current Index to Journals in Education)*—this ERIC database contains bibliographic citations and article summaries to journals. The items in this database are identified by EJ numbers.

7. Internet Search—We might also like to look on the worldwide web for information on our proposed research topic using one of the internet search engines such as Ask Jeeves, LookSmart, Lycos, Netscape, NBCi, Overture, or Google.

ACTIVITY-IDENTIFYING KEYWORDS TO USE IN A LITERATURE REVIEW

Prior to conducting a literature search, we have to identify the keywords or key descriptors that we will be using in our literature search.

1. The relationship between personality correlates and anorexic symptomatology in female undergraduates.

2. The effect of expressive arts on counselling efficacy of adolescents.

3. The effect of cognitive-behavioural techniques on chronic pain management for the elderly.

4. The effect of a structured peer consultation model to address the needs of regular counselling supervision for school counsellors.

5. The relationship between social integration and academic performance among minority university students.

6. The educational characteristics of gifted learning disabled students in the middle school.

Now that you have generated a list of keywords for the above research topics, here are some suggestions, probably similar to those you thought of.

1. The relationship between personality correlates and anorexic symptomatology in female undergraduates. Possible keywords: personality, anorexia, and female.

2. The effect of expressive arts on counselling efficacy of adolescents. Possible keywords: expressive arts, counselling, and adolescent.

3. The effect of cognitive-behavioural techniques on chronic pain management for the elderly. Possible keywords: cognitive, behavioural, chronic pain, and possibly geriatric.

4. The effect of a structured peer consultation model to address the needs of regular counselling supervision for school counsellors. Possible keywords: peer consultation, supervision, and counsellors.

5. The relationship between social integration and academic performance among minority university students. Possible keywords: social integration, academic performance, and minority student.

6. The educational characteristics of gifted learning disabled students in the middle school.

Possible keywords: gifted, learning disabled (or learning disabilities), and possibly middle school or junior high school.

ABSTRACTING THE ARTICLES FOUND IN THE LITERATURE REVIEW

Now that you have selected a research topic, formulated your research problem, selected keywords for a literature search, and identified the articles that potentially will be included in your review of related literature, it is time to read the articles, and write your own summary or abstract of the articles.

The following are procedures for abstracting the contents of your articles.

1. Read the article's abstract or summary to see if it is a usable article for your topic.
2. Skim the entire article making a mental note of the main topics.
3. Write the complete reference in APA style.
4. Classify and code the article according to some system of your own devising. Put the code:
    i. on an index card.
    ii. on the photocopied article (if you photocopied it).
    iii. on the computer so you can sort the article abstracts in any way you wish to.
5. Abstract or summarize the reference by paraphrasing the essential points of the reference. If it's a study you will probably want to include the problem, the procedures, and the major conclusions.
6. Add any thoughts that come to your mind about the article.
7. Indicate any statements that are direct quotations (use quotation marks and also jot down the page number, i.e., Snurd, 1995, p.45). Keep personal reactions separated from direct quotations.

Most student researchers find index cards an indispensable tool for writing, but maybe with the advent of small portable computers, you can take your computer into the library with you and make your notes directly on the computer.

CODING ARTICLES AND PREPARING AN OUTLINE FOR THE REVIEW

By the time you have considered your research problem, read articles about the topic, and abstracted the articles, a tentative outline for your literature review should be coming together. To create an outline at this point:

1. Identify the main points in the order they should be presented.
2. Differentiate each main heading into logical subheadings.
3. Use further subdivisions if necessary.

After you have completed the outline you are ready to start writing your review of the literature.

SUGGESTIONS FOR WRITING A REVIEW OF RELATED LITERATURE

The following are some suggestions that you might find helpful as you start to write your review of related literature.

i.   Make an outline (if you have not already done so).

ii. Analyse each reference in terms of your outline.

iii. Take all references identified for a given subheading and analyse the relationships or differences between them. "Do not present your references as a series of abstracts or annotations. Your task is to organize and summarize the references in a meaningful way." Use APA style for citations to references within your paper. For examples see pages 168 to 174 of the *Publication Manual of the American Psychological Association*.

iv. Your review should start with the articles least related to your research problem and proceed to those most related.

A RESEARCH HYPOTHESIS

A hypothesis is a tentative explanation for certain behaviours, phenomena, or events that have occurred or will occur.

1. The hypothesis states the researcher's expectations concerning the relationship between the variables in the research problem.

2. The hypothesis is a refinement of the research problem. It is the most specific statement of the problem.

3. The hypothesis states what the researcher thinks the outcome of the study will be.

4. The researcher collects data that either supports the hypothesis or does not support it.

5. The hypothesis is formulated following the review of related literature and prior to the execution of the study. The related literature leads the researcher to expect a certain relationship.

6. "A good hypothesis states as clearly and concisely as possible the expected relationship (or difference) between two variables and defines those variables in operational, measurable terms."

7. "A well-stated and defined hypothesis must be (and will be if well-formulated and stated) testable. It should be possible to support or not support the hypothesis by collecting and analysing data."

Formulating a Research Hypothesis

- Hypothesis can be classified in terms of how they were derived
  - inductive hypothesis—a generalization based on observation
  - deductive hypotheses—derived from theory
- A hypothesis can be directional or non-directional.
- Hypotheses can also be stated as research hypotheses (as we have considered them so far) or as statistical hypotheses.
- The statistical hypotheses consist of the null hypothesis (H_0), the hypothesis of no difference and the alternative hypothesis (H_1 or H_A) which is similar in form to the research hypothesis.

To formulate a research hypothesis we start with a research question and

- generate operational definitions for all variables, and
- formulate a research hypothesis keeping in mind
  - expected relationships or differences
  - operational definitions

Formulating Research Hypotheses for Various Types of Research Problems

Let's consider a model or formula for stating a research hypothesis for each of the types of quantitative research types.

1. *A model for stating hypotheses for an experimental or causal-comparative study*   If X is the independent variable, Y is the dependent variable, and S is the Subject, we can state our research hypothesis as Ss who get X do better on Y than subjects who do not get X (or get some other X).

Here is an example: In our study we will hypothesize that first grade girls will show better reading comprehension than first grade boys. Now lets state that as a research hypothesis. It would be a causal-comparative research study.

Research hypothesis   Girls will achieve higher reading comprehension test scores than boys at the end of the first grade.

Operational variables   Reading comprehension will be measured by the Iowa Tests of Educational Development, Reading Comprehension, administered at the end of the year.

Statistical hypotheses   When we get ready to analyse our data we might also wish to state statistical hypotheses for our problem. The statistical hypotheses consist of the null hypothesis (H_0) and the alternative hypothesis (H_1). If we let stand for the mean of the girls and stand for the mean of the boys, our null and alternative hypotheses would be:

In other words the null hypothesis states that there is no difference between the two means on the reading comprehension test scores, while the alternative hypothesis states that the girls mean score on the reading comprehension test will significantly exceed that of the boys. Generally we use a statistical test (e.g. the *t*-test) to decide if the girls score significantly higher than the boys.

2. A model for stating hypotheses for correlational research

If A and B are variables (note that we do not refer to them as independent and dependent variables in correlational research), and C is the subject we can state our research problem as the relationship between A and B for C. (What is the relationship between A and B for C)

We can state our research hypothesis as follows: There will be a significant positive correlation between A and B for C (or significant negative relationship or significant relationship without specifying direction). We will also need to provide operational definitions for A and B and describe C.

Let us analyse the following research example and indicate the research hypotheses.

An instructor investigates the relationship between the number of minutes needed to complete an examination and the score on the examination. He wants to use the data to determine whether there is a significant negative relationship between these two variables.

Research hypothesis   The length of time needed to complete an examination will be negatively correlated with the score on the examination for college students.

FACILITIES AVAILABLE IN INDIA

The National Social Sciences Documentation Centre (NASSDOC) at the CSIR, New Delhi, conducts manual bibliographic searches for a nominal fee. The Indian National Scientific Documentation Centre (INSDOC), New Delhi, has 'online' access to over 300 international databases which cover all the sciences. The National Informatics Centre (NIC), New Delhi, provides online access to MEDLINE using MEDLARS (Medical Literature Analysis Retrieval System). This provides access to one of the world's largest medical databases compiled by the National Library of Medicine at Bethesda, Maryland, USA. It covers approximately 3000 medical journals in 69 languages from all over the world. NIC office in New Delhi is linked to all State capitals and 350 district headquarters via NICNET, a satellite-based communication network.

There are two methods of giving citations:

1. Author prominence, e.g. "Allington (1983) reported that …"
2. Information prominence, e.g. "Incidence of tick infection is more in summer months (Smith, 1975)."
3. Weak Author prominence, e.g. "Several researchers have studied the relationship between bad sanitation and disease occurrence" (Carrington, 2001, Rao *et al.*, 2003 and Whitehead *et al.*, 2006).

Citation form	**Tense**
Information prominent	Simple present
Weak author prominent	Present perfect
General statements	Present perfect
Author prominent	Simple past

Citations can be from distant to close, i.e., from general to specific. Citations can also be arranged chronologically.

16

WRITING THE MATERIALS AND METHODS

In the Materials and Methods, you must give the full details. Most of this section should be written in the past tense. The main purpose of the materials and methods section is to describe the experimental design and then provide enough detail so that a competent worker can repeat the experiments. A general guideline is that you should discuss your methods in sufficient detail that another researcher could take your data and duplicate your results. One of the expectations of performing original research is that someone in the future will do further research on this topic. Such a researcher should be able to use your methodology without having to consult any other source. Most of your readers will skip this section, because they already know (from the Introduction) the general methods you used and they probably have no interest in the experimental details. However, careful writing of this section is critically important because the cornerstone of the scientific method requires that your results, to be of specific merit, must be reproducible and for the results to be adjusted reproducible, you must provide the basis for repetition of the experiments by others.

For materials, include the exact technical specification and quantities and source or method of preparation. Avoiding the use of trade names, and generic or chemical names is usually preferred. Experimental animals, plants and microorganisms should be identified accurately, usually by genus, species and strain designations. The Materials and Methods section is the first section of the paper in which subheadings should be used. When possible, construct subheadings that "match" those to be used in Results. Questions such as "how" and "how much" should be precisely answered by the author and not left for the reviewer or the reader to puzzle over. Statistical analyses are often necessary, but you should feature and discuss the data, not the statistics.

The methods should include the following items:

1. Information to allow the reader to assess the believability of your results.
2. Information needed by another researcher to replicate your experiment.
3. Description of your materials, procedure and theory.
4. Calculations, technique, procedure, equipment, and calibration plots.
5. Limitations, assumptions, and range of validity.

In describing about the method, complete details are to be given for thesis, while for paper publication, the citation of the reference is sufficient. If several alternative methods are used commonly, it is useful to identify the method briefly as well as to cite the reference. For example, it is preferable to state "cells were broken by ultrasonic treatment as previously described" than to state "cells were broken as previously described". When large numbers of microbial strains or mutants are used in a study, prepare strain tables identifying the source and properties of mutants, bacteriophages, plasmids, etc. Do not make the common error of mixing some of the Results in this section. There is only one rule for a properly written Materials and Methods section: Enough information must be given so that the experiments could be reproduced by a competent colleague. A good test, by the way is to give a copy of your finished manuscript to a colleague and ask if he or she can follow the methodology. In Materials and Methods, however, exact and specific items are being dealt with and precise use of English is a must.

The Methods section should be answering the following questions and caveats:

1. Could one accurately replicate the study (for example, all of the optional and adjustable parameters on any sensors or instruments that were used to acquire the data)?
2. Could another researcher accurately find and reoccupy the sampling stations or track lines?

3. Is there enough information provided about any instrument used so that a functionally equivalent instrument could be used to repeat the experiment?

4. If the data is in the public domain, could another researcher lay his or her hands on the identical data set?

5. Could one replicate any laboratory analyses that were used?

6. Could one replicate any statistical analyses?

7. Could another researcher approximately replicate the key algorithms of any computer software?

17

PRESENTING THE RESULTS

Before giving the results, one should give some kind of overall description of the experiments, without repeating the experimental details previously described in Materials and Methods section. Then, the data should be presented. The results should be presented in past tense. Data are mainly presented in Tables and Figures with the minimum of verbal explanation. In the text the same data should not be presented in the form of both Tables and Figures, although there are occasions when this may be justified to make a particular point. Break up your results into logical segments by using subheads.

- The results are actual statements of observations, including statistics, tables and graphs.

- Indicate information on range of variation.

- Mention negative results as well as positive. Do not interpret results—save that for the discussion.

- Lay out the case as for a jury. Present sufficient details so that others can draw their own inferences and construct their own explanations.

- Use S.I. units (m, s, kg, W, etc.) throughout the thesis.

Representative data, rather than endless repetitive data, should be given. If one or only a few determinations are to be presented, they should be treated descriptively in the text. Repetitive determinations should be given in Tables or Graphs. It is important to define even the negative aspects of experiments. It is often good insurance to state that you did not find under the conditions of your experiments. Someone else very likely may find different results under different conditions. If statistics are used to describe the results, they should be meaningful statistics. The results are actual statements of observation, including statistics, tables and graphs.

The results should be short and sweet. Although the results section of a paper is the most important part, it is often the shortest. The results need to be clearly and simply stated, because it is the results that comprise the new knowledge that you are contributing to the world. The earlier parts of the paper (introduction, materials and methods) are designed to tell why and how you got the results; the latter part of the paper (discussion) is designed to tell what they mean. Obviously, therefore, the whole paper or thesis must stand or fall on the basis of the results. Thus, the results must be presented with crystal clarity. Mention both negative as well as positive results. Do not interpret result, save that for discussion.

Do not be guilty of redundancy in the results. The most common fault is the repetition in words or what is already apparent to the reader from examination of the figures and tables. Even worse is the actual presentation, in the text, of all or many of the data shown in the tables and figures. Do not say "it is clearly shown in Table 1 that biocillin inhibited the growth of Brucella". Say "Biocillin inhibited the growth of Brucella (Table 1)".

Where statistical comparisons are needed, use phrases such as "using an analysis of variance, the differences were significant". Full statistical analyses can, again, be included in the Appendices, although full details of the type of statistical analyses performed should be given in the Materials and Methods. Standard errors or standard deviations should be included in Figures and Tables where appropriate.

FORMS OF PRESENTATION OF RESULTS

There are three modes of presentation of data, viz., textual, tabular and graphic. Never present the data in more than one way. The same data should not be presented in both Tables and Figures. The **textual presentation** takes the form of a running composition. Data introduced into a textual discussion may not be in any systematic order; but at convenient points. The data are thus a part of a descriptive discussion presented in paragraph form without comment. Presentation of data is often more emphatic where only a few important facts (e.g. percentages, averages or totals) are to be conveyed. The figures, with few variations, are often clearly understood by those who find difficulty in interpreting facts presented in tabular or graphic forms.

When numerical facts are considerably large and require an effective, clear and consistent view of the whole information, the **tabular form** of presentation is adopted. A table is merely a repository of numerical facts, which makes it possible for the researcher to present a huge mass of data in a detailed and orderly manner within minimum space. A tabular presentation is considered the cornerstone of interpretation in research work because it simplifies the complex data, facilitates comparisons, reveals patterns and gives identity to the formation at a glance. Do not construct a table unless repetitive data must be presented. Cost of publishing tables is very high compared with that of text. If you have a few determinations, give the data in the text. Tables should not be constructed with the word list.

Make sure that your variables are in different columns. Your rows for any given column should represent different observations of a given variable. In designing a Table, the value to be conveyed should be in vertical column, as the eye is able to grasp much better in a vertical form. The vertical column is called the box head; the horizontal column is called the "stub head" and the portion containing the data is called the field. The table should always be introduced. It should follow as closely after its first mention as possible, e.g. mean, standard deviation and standard error of means may be found in Table 20.

A good general rule to follow about the placement of tables and figures is that, if they occupy a whole page or more, they should be presented on a new page. If they occupy less than half a page, they may be surrounded by text or may be given in a separate page. In practice, it is often more convenient to have all tables and figures on separate pages because of the time saved in the event of retyping. If the data in the table is essential for the conclusion which follows, it should be presented in the main body of the text. If however, the material is basic test data, then its proper place is in an Appendix.

A general table contains at least the following important parts, viz., table number, title of the table, stubs (designations of the row headings), body of the table and footnote. The table should suit the size of the paper being used and it should contain more rows than columns. The units of measurement should be given clearly. The title of a Table is a description of the contents of the table. A complete title has to answer the questions what, where and when in that sequence.

1. What precisely are the data in the table?
2. Where the data occurred?
3. When the data occurred?

All tables should be numbered to permit easy identification. The usual practice is to use Arabic numbers (1, 2, 3, etc.). The word TABLE in capitals and its appropriate number is centred on the page. It is placed above the table, one double space above the title of the table. If a table appears on a page with text, it is separated from the text by a triple space above and below. The same rule applies for numbering of Figures, except that the word FIGURE in capitals with its number is placed one double space below the Figure. One double space below is the title of the Figure.

There are two forms of tabulation: i) informative and ii) interpretative. When data is systematically compiled in tables, with a view to merely providing a convenient means of presentation for easy reference, and further use, it is **informative tabulation**. There is no intention of presenting comparisons, relationships or significance of figures. On the other hand, when a significant aspect of data are to be presented, incorporating full information to facilitate a proper consideration of all related facts, **interpretative tabulation** is adopted. Interpretative tables show comparative figures, whether absolute or percentage, or average of various kinds and other forms of comparison between related facts.

An informative table showing the payment of daily wages to workers in an establishment is as follows.

Date	Department	No. of workers	Basic wages	Overtime	Total

A simple interpretative table showing a dependent variable against an independent variable is shown below.

Period	No. of workers
First week	2250
Second week	2010
Third week	1908
Fourth week	2390

A complex interpretative table showing the relationship of more than one group of items or dependent variables tabulated against an independent variable is given below.

Period	Men	Women	Total	Percentage of total
First week	1450	800	2250	26.29
Second week	1390	620	2010	23.49
Third week	1170	738	1908	22.29
Fourth week	2020	370	2390	27.93
Total	6030	2528	8558	100.00

Another method of classification of table is:

1. Simple and complex tables
2. General-purpose table
3. Special-purpose (or summary) table

Simple and Complex Tables

The distinction between simple and complex tables is based upon the number of characteristics studied. In a simple table only one characteristic is shown. Hence, this type of table is also known as one-way table. For example, the number of trainees in a typing institute can be given as a one-way table.

Age group (in years)	No. of trainees
Below 20	35
20 to 30	42
30 to 40	16
Above 40	7

In the complex table, two or more characteristics are shown. Such tables are more popular in practice because they enable full information to be incorporated and facilitate a proper consideration of all related facts. When two characteristics are shown, such a table is known as a two-way table. For example, the number of trainees in the typing institute in different age groups according to sex can be given in a complex table as follows.

Age group (in years)	Male	Female	Total
Below 20	19	16	25
20–30	22	20	42
30–40	11	5	16
Above 40	5	2	7
Total	57	43	100

General-Purpose Table

General-purpose or reference tables contain information which serve an archival function and often need to be laid out for economy of space, while preserving data accurately. It is extremely important that they include good meta-data, the descriptive information which allow the data to be correctly interpreted—usually a comprehensive version of the "what, where and when". Good tables should be easy to read across rows and down columns, easy to understand, and easy to refer to in the text of your report. They should also include only relevant data from the results.

An example of general-purpose table is the textual (word) table.

Textual (word) table   Often research data presentation need tables that have textual data in the body. Usually this is the case when qualitative data are dealt with. These tables serve the same function as any table, to make comparisons of items easy. These tables are also used when you want to present examples, which may be grouped in a certain way, or when you want to show categories of different items.

Special-Purpose (or Summary) Tables

Special tables are intended to be assimilated quickly by the reader or viewer. They condense a large mass of data and bring out the distinct pattern in a data in an attractive form. It enables comparison to be made easily among classes of data.

Examples of special tables are statistical tables, numerical tables and two-way tables.

Statistical tables   These tables can present descriptive or inferential statistics or both. Descriptive statistics are tabulations such as mean, standard deviation, mode, range, or frequency. Inferential statistics refers to statistical tests. In such tables, statistical test values are summarized.

Numerical tables   These are the most common types of data, which typically represent quantitative data, but sometimes may present a combination of quantitative and qualitative data. As its name suggests, most of the body of the table consists of the summation of specific number values.

Two-way table   Also called a "contingency table" it is a useful tool for examining relationships between categorical variables. The entries in the cells of a two-way table can be frequency counts or relative frequencies.

DIAGRAMMATIC AND GRAPHIC PRESENTATION

A graph should be reserved for exciting findings or interesting, but unexpected, results. Trends, departures from trends, dramatic behaviours of variables, etc., are good candidates for graphs. The diagrammatic and graphic form of presentation is preferred where significant relationship of data are to be rendered visible at a glance. Graphs and diagrams have a visual appeal and therefore prove to be more impressive to a layman. The advantages are:

1. They give a bird's-eye view of the entire data and therefore, the information presented is easily understood.

2. They are attractive to the eyes. Figures are dry, but diagrams delight the eyes. For this reason, diagrams create greater interest than cold figures.

3. They have great memorizing effect. The impressions created by diagrams last much longer than those created by the figures presented in a tabular form.

4. They facilitate comparison of data relating to different periods of time or different regions. Diagrams help one in making quick

and accurate comparison of data. They bring out hidden facts and relationship and can stimulate as well as aid analytical thinking and investigation.

There are several types of diagrams used for displaying facts. Those which are more frequently used are given below:

1. One-dimensional diagrams, e.g. bar diagrams.
2. Two-dimensional diagrams, e.g. rectangles, squares and circles.
3. Three-dimensional diagrams, e.g. cubes, cylinders and spheres.
4. Pictograms and cartograms.

The choice of diagram primarily depends on i) the nature of data and ii) the type of people for whom the diagram is meant.

One-Dimensional Diagram or Bar Diagram

Bar diagrams are the most common type of diagrams used in practice. A bar is a thick line whose width is shown merely for attention. They are called one-dimensional because it is only the length of the bar that matters and not the width. Advantages of bar diagrams are i) they are readily understood even by those unaccustomed to reading charts or those who are not chart-minded; ii) simplest and the easiest to make and iii) when a large number of items are to be compared, they are the only form that can be used effectively.

While constructing the bar diagrams, the following points should be kept in mind. The width of the bars should be uniform throughout the diagram. The gap between one bar and another should be uniform throughout. Bars may be either horizontal or vertical. The vertical bars should be preferred because they give a better look and also facilitate comparison. While constructing the bar diagram, it is desirable to write the respective figures at the end of each bar so that the reader can know the precise value without looking at the scale.

The types of bar diagrams are

1. Simple bar diagram
2. Sub-divided bar diagram—each bar is subdivided into its various components.
3. Multiple bar diagram—two or more sets of interrelated data are represented.

4. Percentage bar diagram—to show relative changes in data. The length of the bar equals 100.

5. Deviation bars used to represent net quantities, excess or deficit, e.g. net profit, loss, etc. Have both positive and negative values.

Simple bar diagram

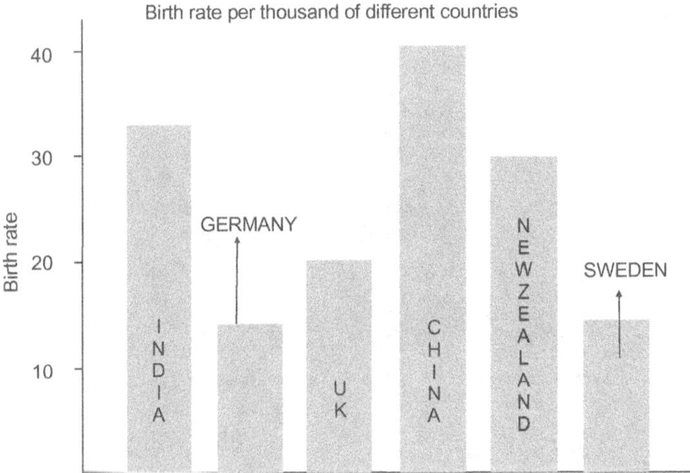

Sub-divided bar diagram

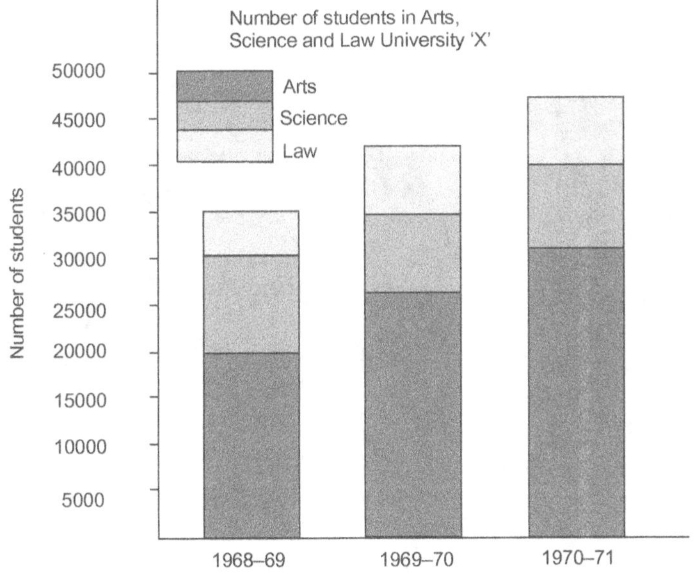

Multiple bar diagram

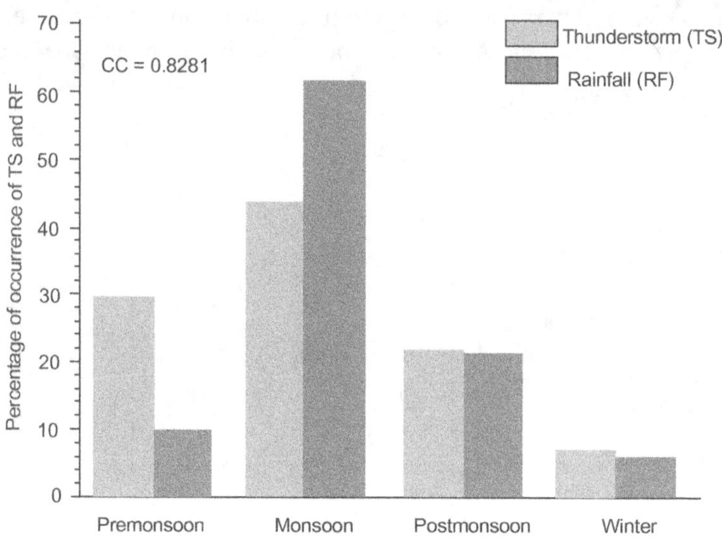

Percentage bar diagram

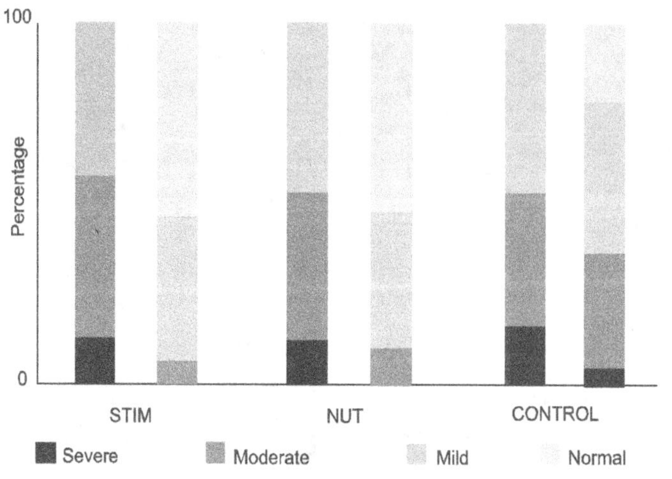

Deviation bars

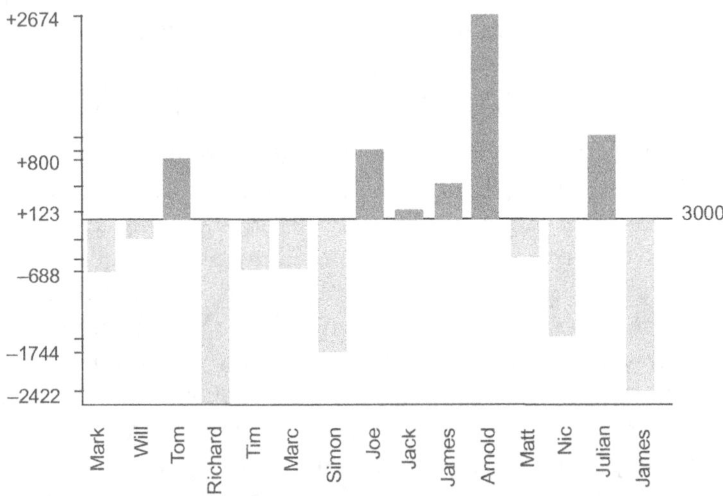

The difference to a normal bar diagram is that it does not show absolute values but the deviation from a base value.

Horizontal bar diagram

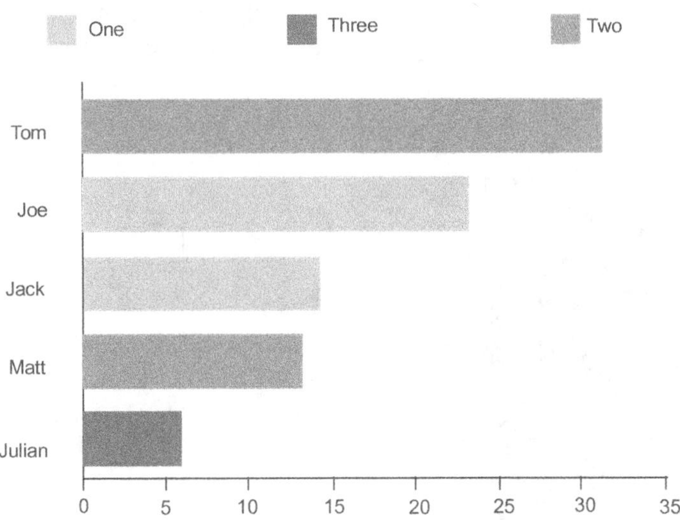

Horizontal bar diagrams can be useful when the order of the bars is important (for example if we have a rank order), or when the identifiers for the bars are too long and we do not want them to be rotated.

Two-Dimensional Diagrams

In this diagram, the length as well as the width of the bars are considered. Thus, the area of the bars represents the given data. Such two-dimensional diagrams are also called as surface diagrams or area diagrams. The types of two-dimensional diagrams are 1) rectangles, 2) squares and 3) circles. **Rectangles** are quite popular for presentation. Since the area of a rectangle is equal to the product of its length and width, while constructing such a diagram, both length and width are considered. We may represent the figures as they are given or may convert them to percentage and then subdivide the length into various compartments. The rectangular method of diagrammatic presentation is difficult to use where the value of the item varies widely. For drawing a **square diagram**, one has to take the square-root of the values of various items that are to be shown in the diagrams, then select a suitable scale to draw the squares. **Circles** can be used in all those cases in which squares are used. Circles are compared on an area basis rather than on a diameter basis. Circles are difficult to compare and not popular. Pie diagrams are very popularly used to show percentage breakdowns. With the help of pie diagram, we can show how the

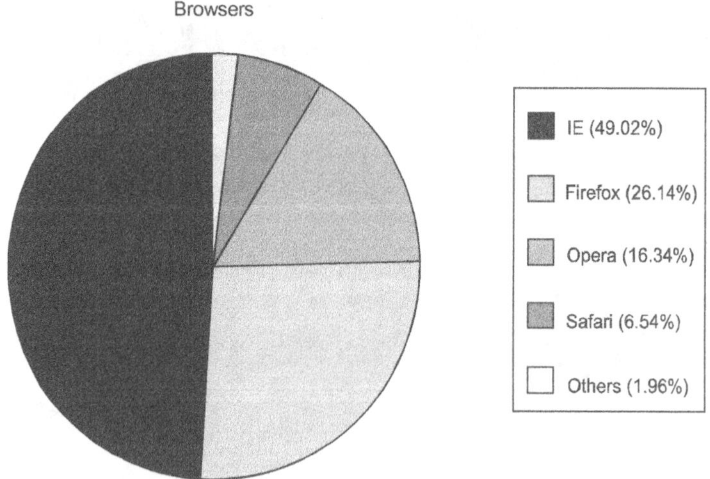

expenditure of the Government is distributed over different heads like Agriculture, Irrigation, Industry, Transport, Defence, etc. The pie chart is so called because the entire graph looks like a pie, and the components resemble slices cut from pie. While making comparisons, pie diagrams should be used on a percentage basis and not on an absolute basis.

Three-Dimensional Diagrams

Three-dimensional diagrams, also known as volume diagrams, consist of cubes, cylinders, spheres, etc. In such diagrams, three aspects, viz., length, width and height have to be taken into account. Such diagrams are used where the range, i.e., the difference between the smallest and the largest value is very large.

Pictographs and Cartograms

Pictographs are popularly used in presenting statistical data. They are not abstract presentations such as lines or bars, but really depict the kind of data we are dealing with. Pictures are attractive and easy to comprehend and as such this method is particularly useful in presenting statistics to the layman. When pictographs are used, data are represented through pictorial symbols that are carefully selected. Compared with other types of diagrams, pictographs have a greater attraction value, and therefore, where the attention of masses is to be drawn such as in exhibition and fairs, they are very popularly used. Facts portrayed in pictorial form are generally remembered longer than facts presented in tables or in non-pictorial charts. Limitations are they are difficult to construct and they give only the overall picture, not the minute details.

Cartograms or statistical maps are used to give quantitative information on a geographical basis. They are thus used to represent spatial distributions. Statistical maps should be used only where geographical comparisons are of primary importance and where approximate measures will suffice. Diagrams give only an approximate idea of the measurement of the phenomena. Therefore, small differences in large measurements cannot be shown. Diagrams can show at best one or two types of comparisons. They prove unsuccessful when multiple comparisons are required to be shown especially when the difference between the two measurements is very large and thus diagrams are not capable of further analysis.

GRAPHS

Graphs are used to illustrate trends and relationships and get the reader to visualize and interpret the data more readily. The fluctuations occurring over a period of time can be projected more emphatically by graphs than a table. Graphs can be used for comparison of two categories of results. Graphs help to express a general trend while tables express a factual one. Facts and figures can also be displayed through graphs. Graphs are generally used for representing continuous series and time series as these are more accurate and also mathematically precise. For preparing graphs, two simple lines which interact with each other at right angles are drawn. The lines are known as co-ordinate axes. The point of intersection is called point of origin or xero point. The horizontal line is known as the axis of x or "abscissa" and the vertical line the axis of T or "ordinate". The independent variables are indicated along the X-axis and dependent variables are shown in Y-axis. The scales chosen must be appropriate such that the graphs are compact. The graphs are divided into two types:

1. Time series
2. Frequency distribution

Time Series

When we observe the values of a variable at different points of time, the series so formed is known as time series. When a series indicates a change on the basis of time, it is known as time series. In time series, time is shown on horizontal line, while values are shown on vertical line. Various points are joined together to show continuity. These are the simplest to understand, easiest to make and most adaptable to many uses. On the X-axis we take the time and on the Y-axis the value of the variable and join the various points by straight lines. The graph so formed is line graph.

Types of Graphs

Time graph   This type of graph is widely used. This is used to show the relationship between widely different parameters.

Silhoutte graph   This is used to emphasize variations above and below the standard. This is also known as scatter graph and is very popular.

Band graph   This type is used to show a total variable and also the components such as the cost of producing milk, divided by the cost of feed, cost of management, etc. over a period of time. The band graph shows how and in what proportion the individual items comprising the aggregates are distributed. The various component parts are plotted one over the other. This is useful in dividing total costs, into component cost, total sales into department or district or individual salesman's sale, etc.

Histograph   This is used to show the frequency distribution such as different yields of cows.

Pie graph   This is used to represent differences in volume. This represents relative proportions of the figures.

Frequency Series

The main types of frequency series that are commonly represented through graphs are 1) histogram, 2) frequency polygon, 3) frequency curve and 4) ogives or cumulative frequency curves.

Histograph   A histogram is drawn to represent relative frequency size of different groups. In a histograph, the variable is always taken on the X-axis and the frequencies depending on it on the Y-axis. Each class is represented by a rectangle which is proportional to its class interval. A histogram is a set of vertical bars whose areas are proportional to the frequencies represented. The distinction between a histogram and bar diagram lies in the fact that whereas bar diagram is one-dimensional, a histogram is two-dimensional. In other words, in a bar diagram, only the length of the bar is material and not the width, whereas in a histogram both the length as well as width are important. The histogram is commonly used for graphical presentation of a frequency distribution.

Frequency polygon   This is a graph of frequency distribution. It displays the same aspect of data as histogram. In fact, the areas of both are identical. The only difference between the two is that histogram depicts each frequency group separately whereas the polygon does it collectively. To construct a polygon, we can draw a histogram of the frequency distribution and mark the midpoints of the tops of all the rectangles and connect them with the help of straight lines. This will show a curve with sharp corners at each joint.

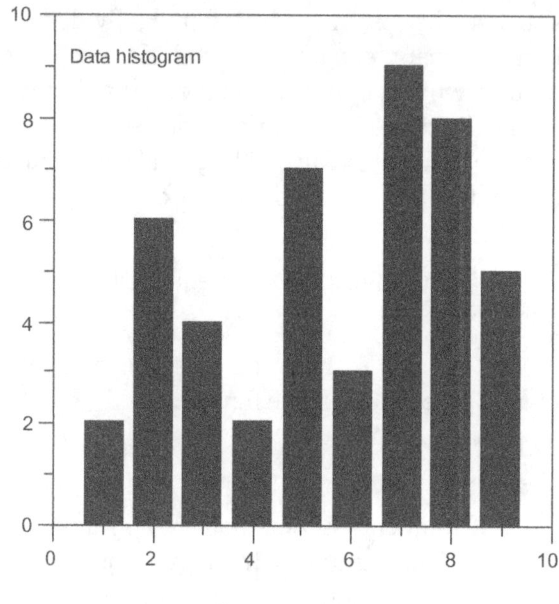

10
Data histogram
8
6
4
2
0
0   2   4   6   8   10
Runs scored

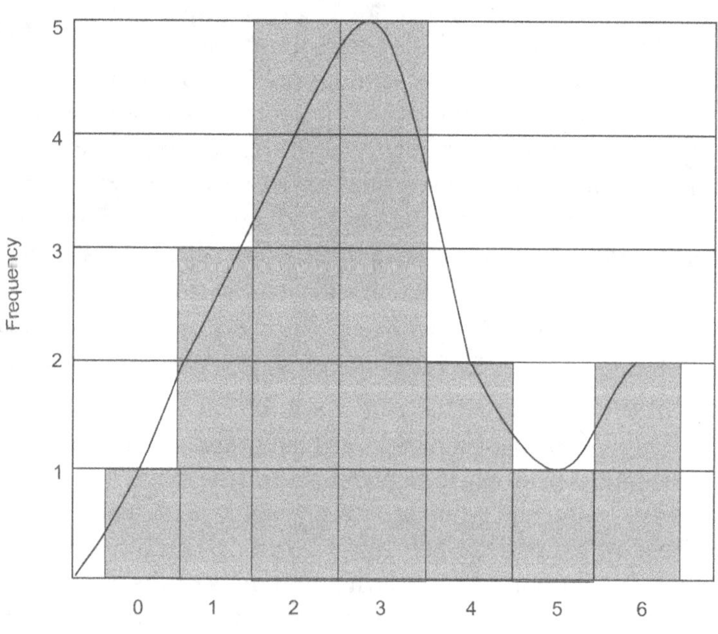

5
4
3
Frequency
2
1
0   1   2   3   4   5   6
Runs

Frequency curve   A smooth frequency curve can be drawn when ruggedness of the polygon is smothered out and a bell-shaped curve is evolved free hand in such a way that the area included under the curve is approximately the same as that of the polygon. The object of this is to eliminate as far as possible accidental variations that might be present in the data.

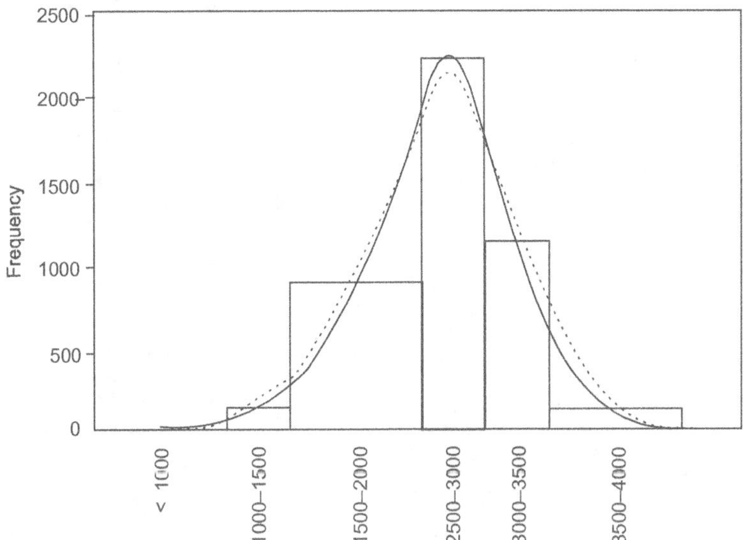

Ogives   Also known as cumulative frequency curve, ogives are basically drawn to show cumulative rise or fall in the frequencies. Ogive is of two types: In the "less than" type, the curve has a rising trend, whereas in the "more than" type, it has a falling trend. In the "less than" method, we start with the upper limits of the classes and go on adding the frequencies. When these frequencies are added, we get a rising curve. In the "more than" method, we start with the lower limits of the classes and from the frequencies, we subtract the frequencies of each class. When these frequencies are plotted we get a declining curve. For determining the frequencies of the cumulative frequency curve, various frequencies are cumulatively added from top to bottom and bottom to top in the case of "less than" and "more than" type, respectively as illustrated below:

Mark distribution in a class of 70 students

Marks	Frequency	Less than	More than
0–20	2	2	68
21–30	3	5	65
31–40	10	15	55
41–50	20	35	35
51–60	25	60	10
61–70	5	65	5
71–80	3	68	2
81–90	2	70	0

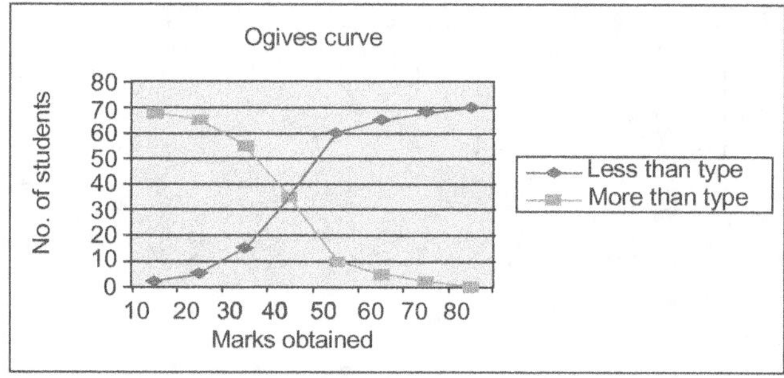

These presentations are useful to determine as well as to portray the number of cases above and below a given value. This is used to compare two or more frequency distribution.

18

WRITING THE DISCUSSIONS

Discussion is usually the hardest section to write. And, whether you know it or not, many papers are rejected by journal editors because of a faulty discussion, even though the data of the paper might be both valid and interesting. Start with a few sentences that summarize the most important results. The discussion section should be a brief essay in itself, answering the following questions:

1. What are the major patterns in the observations? (Refer to spatial and temporal variations.)

2. What are the relationships, trends and generalizations among the results?

3. What are the exceptions to these patterns or generalizations?

4. What are the likely causes (mechanisms) underlying these patterns resulting in predictions?

5. Is there agreement or disagreement with previous work?

6. Interpret results in terms of background laid out in the introduction— what is the relationship of the present results to the original question?

7. What is the implication of the present results for other unanswered questions in earth sciences?

8. *Multiple hypotheses* There are usually several possible explanations for results. Be careful to consider all of these rather than simply pushing your favourite one. If you can eliminate all but one, that is great, but often that is not possible with the data in hand. In that case you should give even treatment to the remaining possibilities, and try to indicate ways in which future work may lead to their discrimination.

9. Avoid bandwagons. Avoid jumping a currently fashionable point of view unless your results really do strongly support them.

10. What are the things we now know or understand that we didn't know or understand before the present work?

11. Include the evidence or line of reasoning supporting each interpretation.

12. What is the significance of the present results?

Discussion presents your opinion, while the results present the facts. The first sentence in the discussion should briefly describe the most important result you obtained. Explain why the result is important and what it has added to our knowledge or understanding of the phenomenon under consideration.

Discussion should assess how far the objectives are achieved. This should also give a summary of what you have done. There should be critical remarks on the findings in the context of the existing literature. It should give the impact of the work. Difficulties faced should be emphasized and scope for further work should be mentioned. Do not repeat what you have said earlier. You can remind few of the earlier results like "as pointed out earlier".

COMPONENTS OF DISCUSSION

1. Try to present the principles, relationship and generalization shown by the Results. And bear in mind, in a good discussion, you discuss—you do not recapitulate the results.

2. Point out any exceptions or any lack of correlation and define unsettled points. Never take the high-risk alternative of trying to cover up or fudge data that do not quite fit.

3. Show how your results and interpretations agree (or contrast) with previously published work.

4. Don't be shy; discuss the theoretical implications of your work, as well as any possible practical applications.

5. State the things we now know or understand that we did not know or understand before the present work.

6. State the significance of the present results.

7. State your conclusions as clearly as possible.

8. Summarize your evidence for each conclusion.

9. Highlight possible areas for further research.

We arrive at one of the three conclusions:

1. Our conclusions agree with what everyone already knows, i.e., supports contemporary knowledge.

2. Our conclusions sustain ideas and views suspected to be true for which there was previously little corroborative evidence.

3. Our conclusions are entirely new and have not been thought of before.

In the discussion, your verb tense should swing back and forth between present and past. Other people's work (established knowledge) should be described in the present tense, but your own results should be described in the past tense.

Primary information elements (Specific reference to the study)

1. Reference to the main purpose or hypothesis of the study

2. Review of the most important findings

3. Possible explanations for or speculations about the findings

4. Limitations of the study that restrict the extent to which the findings can be generalized

Later information elements (General statements about the study)

1. Implications of the study (generalizations from the results)

2. Recommendations for future research and practical applications.

Researcher's position on information

Position	Information element
One possible explanation is We can no longer assume	that speed jobs do not tax older workers
We acknowledge clearly	that this technique has promise as a tool in evaluation of forages

Expressions for restating the hypothesis

Main clause	Noun clause
It was anticipated	
The theory leads us to infer	that older workers in speed jobs would have poorer performance than younger workers.
In line with this hypothesis, we assumed	
The results seem inconsistent with our hypothesis	

Expressions for explaining findings

These results can be explained by assuming that	
One reason could be that	skill increases with experience.
It is unlikely that	

Expressions for suggesting implications

	suggest
	imply
These findings	lend support to the assumption that
	led us to believe
	provide evidence

Too often, the significance of the results is not discussed or not discussed adequately. The discussion should end with a short summary or conclusion regarding the significance of the work. The primary purpose of the Discussion is to show the relationships among observed facts. To emphasize this point, a story was told about the biologist who trained a flea.

After training the flea for many months, the biologist was able to get a response to certain commands. The most gratifying of the experiments was the one in which the Professor would shout the command "jump", and the flea would leap into the air each time the command was given. The Professor was about to submit this remarkable feat to posterity via a scientific journal, but he, in the manner of the true scientist, decided to take his experiment one step further. He sought to determine the "location of the receptor organ" involved. In one experiment, he removed the legs of the flea, one at a time. The flea obligingly continued to jump upon command, but as each successive leg was removed, its jump became less spectacular. Finally, with the removal of its last leg, the flea remained motionless. Time after time, the command failed to get the usual response. The Professor decided that at last he could publish his findings. He set pen to paper and described in meticulous detail the experiments executed over the preceding months. His conclusion was one intended to startle the scientific world: "When the legs of a flea are removed, the flea can no longer hear".

A science teacher set up a simple experiment to show her class the danger of alcohol. She set up two glasses, one containing water, and the other containing gin. Into each she dropped a worm. The worm in the water swam merrily around. The worm in the gin quickly died. What does this experiment prove she asked. Little Johnny from the back row piped up: "It proves that if you drink gin, you wouldn't have worms".

Discussion should end with **conclusions**. This conclusion section usually consists of a concise recapitulation of the objectives, results obtained and conclusions and possible future research objectives. The following are the points to be covered in the conclusions.

1. What is the strongest and most important statement that you can make from your observations?

2. Refer back to the problem posed, and describe the conclusions that you reached from carrying out this investigation, summarize new observations, new interpretations and new insights that have resulted from the present work.

3. Include the broader implications of your results.

4. Do not repeat word by word the abstract, introduction or discussion.

This is where you say why you think your story is a good one and present evidence from your work to support your claim. The fate of your hypothesis is revealed here: did it stand, fall, or require modification? You may briefly compare your work with that of others, present whatever new knowledge has been gained from your work, and suggest what may be done to further new knowledge. The Conclusions should give a sense of fulfilment and finality to your thesis, and give the reader some satisfaction that the time spent on reading it has not been in vain.

Recommendations for further research may be suggested at the end of discussion including the following points.

◎ Remedial action to solve the problem.

◎ Further research to fill in gaps in our understanding.

◎ Directions for future investigations on this or related topics.

19

CITING THE REFERENCES

The bibliography is usually placed immediately after the last chapter of the thesis, but some writers prefer to place the bibliography after the appendix. Every book, article, thesis, document or manuscript, which has been read or examined or cited should be included in the list of references. Pagination is continuous and follows the page numbers in the text. The term References is generally used for written assignments and papers for publications, whereas in thesis writing, the term Bibliography should be used as it is supposed to give complete particulars of the references. There are two rules to follow in the Reference section. First, you should list only significant, published references. Second, check all parts of every reference against the original publication before the manuscript is submitted and perhaps again at the galley proof stage.

REFERENCE STYLES

Journals vary considerably in their style of handling references. One person looked at 52 journals and found 33 different styles for listing references. Some journals print titles of articles and some do not; some insist on inclusive pagination, whereas others print first pages only. The smart author writes out references in full. Thus in preparing a manuscript, he or she has all the needed information. It is easy to edit out information; it is indeed laborious to track down references to add article titles or to edit pages when you are required to do. Even if you know that the journal to which you plan to submit your manuscript uses a short form, you should still be wise to establish your reference list in the complete form. This is good practice because i) the journal you selected may reject your manuscript, resulting in your decision to submit the manuscript to another journal and ii) it is more than likely that you will use some of the same references again, in later research papers, review articles or books. When you submit a manuscript for publication, make sure that the references are presented according to the instructions to authors.

Although there is an almost infinite variety of reference styles, most journals cite references in one of the three general ways, that are usually referred to as "name and year" (Harvard system of reference), "number from alphabetical list" (Alphabet-number system) and "number in order of citation" (Citation order system).

Harvard System of References (Name and Year System)

This system is very popular and is still used in many journals. Its big advantage is its convenience to the author. Because the references are unnumbered, references can be added or deleted easily. If there are two or more "Smith and Jones (1950)" references, the problem is easily handled by listing the first as "Smith and Jones (1950a)" and the second as "Smith and Jones (1950b)", etc. This is inconvenient to the reader when (often in the Introduction) a large number of references must be cited within one sentence or paragraph. The inconvenience to the publisher is obvious: increased costs. When "Smith, Jones and Higginbotham (1975)" can be converted to "(7)", composition (typesetting) and printing costs can be reduced. Usually, for referring to more than two authors, *et al.* is used. Author(s), year of publication, title of the article, name of the journal, volume and issue number of journal, and page numbers are written in this order. Some people put year after the page numbers, e.g. Anderson, S.A. Studies on coccidian in chicken. *Ind. Vet. J.* 15: 116–125. 1978.

Alphabet-Number System

This system, citation of number from an alphabetized list of references, is a modern modification of the name and year system. Citation by numbers keeps printing expenses within bounds; the alphabetized list, particularly if it is a long list, is relatively easy for authors to prepare and readers (especially librarians) to use. Some people claim that citation of numbers cheats the reader. The reader should be told the name of the person associated with the cited phenomenon. As you cite references in the text, decide whether names or dates are important. If you want to feature the name of the author, do it within the context of the sentence: "The role of carotid sinus in the regulation of respiration was discovered by Smith" (13). If you want to feature the date, you can also do that within the sentence: "Streptomycin was first used in the treatment of tuberculosis in 1945" (45).

Citation Order System

This system is simply citing the references (by number) in the order that they appear in the paper. This system avoids the substantial printing expense of the name and year system, and readers often like it because they can quickly refer to the references if they so desire in one-two-three order as they come to them in the text. This system is good for short papers. For long papers, with many references, citation order is probably not a good system. It is not good for the author, because of the substantial numbering chore that results from addition or deletion of references.

Titles and inclusive pages   Should article titles be given in references? Normally, you will have to follow the style of the journal. It is recommended that one should give complete references. By denoting the overall subjects, the article titles make it easy for interested readers to decide whether they need to consult none, some or all of the cited references. The use of inclusive pagination (first and last page numbers) makes it easy for the potential users to distinguish between one page notes and 50 page review articles.

Journal abbreviations   Although journal styles vary widely, one aspect of reference citation has been standardized in recent years, i.e., journal abbreviations. By noting a few of the rules, authors can abbreviate many journal titles, even unfamiliar ones. It is helpful to know, for example, that all "-ology" words are abbreviated at the "l."

Bacteriology	*Bacteriol.*
Physiology	*Physiol.*
Journal	*J.*
Biological	*Biol.*
Veterinary	*Vet.*
Proceedings	*Proc.*
Natural	*Nat.*
Academy	*Acad.*

An exception to be remembered is that one-word titles (*Science, Biochemistry, Nature*, etc.) are never abbreviated.

The bibliography should follow a logical arrangement in alphabetical order. The surname starts with the left margin so that it stands out clearly. Second and subsequent lines of the same entry are single-spaced starting 3 spaces in from the left margin. Double-spacing is allowed between entries.

20

USING THE FOOTNOTES

Footnotes are conventional validity and explanatory procedures which should be used sparingly and only when the material being presented clearly needs amplification or acknowledgement. Footnotes should appear only in the body of the paper or thesis, never in an abstract. As the name implies, footnotes are usually found at the foot of a page although in some manuscripts they appear at the end of each chapter or at the end of the paper.

USE OF FOOTNOTES

Footnotes are commonly used to

1. validate a point, statement or argument;
2. explain, supplement or amplify material that is included in the main body of a paper;
3. provide cross-references to other sections of a paper;
4. acknowledge a direct quotation or indirect quotation and
5. provide the readers with sufficient information to enable them to consult sources independently.

The following information is usually included in footnotes:

1. source of information, usually the name of the author;
2. title of the source;
3. exact page(s) of the source of reference;
4. date of publication and
5. publisher and place of publication (optional).

PLACEMENT OF FOOTNOTES

Footnotes are usually placed at the foot of a page. But they may be variably placed at the end of the chapter or at the end of the paper when they may be called as endnotes.

Reference to footnotes wherever they may be placed is made by the use of superscripts in the body of the text where the particular reference is given. When footnotes are placed at the foot of the page, they are separated from the text by a fifteen-space solid line drawn from the left-hand margin and one double space below the last line of the text. When footnotes are placed at the end of a chapter or paper, a centred heading FOOTNOTES is required.

FORMAT OF FOOTNOTES

The recommended practice is to indent the first line of the footnote in the same way as paragraphs. Footnotes occupying more than one line are single-spaced and only the first line is indented. A double space separates successive footnotes. Footnotes are usually numbered consecutively throughout a chapter or continuously through a paper or thesis.

CONVENTIONS IN FOOTNOTING

There are a number of conventions used in footnoting:

1. In the first footnote referring to each source, it is usual to give the full name of the author in its normal order (that is first name or initial and second initial preceding surname).

2. In citing the reference details, bibliographical procedures are followed: titles of complete works are underlined; names of articles and similar material are enclosed by double quotation marks;

3. After the first reference is spelled out in a footnote, it is not necessary to repeat the name of the author, publisher and so on. There are accepted abbreviations which avoid repetition and lengthy documentation.

i. *Ibid.* (*ibidem*—in the same place or work). If reference is made to a different page of a source supplied immediately above, it is possible to use the term Ibid. It is used when 2 or more successive footnotes refer to the same work, e.g. [2]*Ibid.*, p. 147.

ii. *loc. cit.* (*loco citato*—in the place cited). If reference is made to the same page as a preceding, but not immediately preceding reference, the last name of the author and the phrase *loc.cit.* are used, e.g. [3]Hudson, *loc. cit.* p.150.

iii. *op.cit.* (*opera citato*—in the work cited). If reference is made to the same work as a preceding, but not immediately preceding reference, *op. cit.* precedes page reference but follows author's name, e.g. [7]Poole, *op. cit.* p. 238.

iv. The abbreviation p. for page and pp. for pages is the acceptable method of citing page reference.

SOME POINTS TO REMEMBER

1. Decide whether a footnote strengthens or validates a point in your paper.

2. Include footnotes in your first draft.

3. Check each footnote for accuracy and for correct format.

4. Having adopted a method of footnoting, be consistent throughout the whole assignment or thesis.

5. Footnotes should be concise, but clarity and readability should never be sacrificed for brevity.

6. All footnotes are single-spaced, no matter where they appear on a manuscript, but are separated by a double space.

7. All footnotes regardless of length are terminated by a full stop.

8. The same bottom margin should be maintained on each page of the typescript, regardless of the number of footnotes.

9. A footnote may be continued on consecutive pages, but where a footnote is very long, an assessment should be made as to whether an appendix might be more appropriate than a footnote.

21

LISTING OF AUTHORS
AND ADDRESSES

The easiest part of preparing a scientific paper is simply the entering of the byline: the authors' addresses. Often, the Head of the laboratory is placed last (second of two authors, third of three, etc.). As a result, the terminal spot becomes a choice placement for its presumed prestige. The modern tendency has been to define the first author as the senior author—the primary progenitor of the work being reported.

DEFINITION OF AUTHORSHIP

Perhaps we can now define authorship by saying that the listing of authors should include those, and only those, who actively contributed to the overall design and execution of the experiments. Further, the authors should normally be listed in order of importance to the experiments, the first author being acknowledged as the senior author, the second author being the primary associate, the third author possibly being equivalent to the second, but more likely having a lesser involvement with the work reported. An author of a paper should be defined as one who takes intellectual responsibility for the research results being reported. Each listed author should have made important contribution to the study being reported. The scientific paper should list as authors only those who contributed to the work.

DEFINING THE ORDER

Suppose that Scientist A designs a series of experiments that result in important new knowledge and then Scientist A instructs Technician B exactly how to perform the experiments. If the experiments work out and a manuscript results Scientist A should be the sole author, even though Technician B did all the work. Of course, the assistance of Technician B should be recognized in the Acknowledgments. Now let us suppose that the above experiment did not work out. Technician B takes the negative results to Scientist A and says something like, "I think we might get this strain to grow if we change the incubation temperature from 24 to 37°C and if we add serum albumin to the medium". Scientist A agrees to a trial, the experiments this time yield the desired outcome and a paper results. In this case, Scientist A and Technician B, in that order, should be listed as authors.

LISTING THE ADDRESSES

With one author, one address is given (the name and address of the laboratory in which the work was done). If, before publication, the author has moved to a different address, the new address should be indicated in a "Present Address" footnote. When two or more authors are listed, each in a different institution, the addresses should be listed in the same order as the authors.

Remember that an address serves two purposes. It serves to identify the author; it also supplies the author's mailing address. The mailing address is necessary for many reasons, the most common one being to denote the source of reprints.

22

USING QUOTATIONS

During the research process, particularly during note-taking, the research student may copy extracts from sources verbatim with the intention that these extracts may be incorporated into the written research report. Although in the initial note-taking period, it is wise to be provided with an abundance of appropriate possible quotations from which to choose, in the writing of the final report, it is essential that quotations should be judiciously selected and sparingly used.

WHEN TO QUOTE

Although the final decision of when to quote depends on the problem being investigated and on the judgement of the research student, there are a number of guidelines which can assist the student in reaching a decision.

1. Direct quotations would be used only when the original words of the author are expressed so concisely and convincingly that the student cannot improve on these words.

2. Direct quotations may be used for documentation of a major argument where a footnote would not suffice. In this case, quotation should be limited in length and comprise only essential passages.

3. Direct quotations may be used when the student wishes to comment upon, refute or analyse ideas expressed by another writer.

4. Direct quotations may be used when changes, through paraphrasing, might cause misunderstanding or misinterpretation, for instance in citing the words of the law, in stating assumptions underlying a statistical procedure or quoting from government publications.

5. Direct quotations should be used when citing mathematical, scientific and other formulae.

WHAT TO QUOTE

Although there is some flexibility permitted in deciding "when" to quote, there are more stringent conventions stating what should be quoted.

1. The exact words of an author or the exact words from an official publication must be quoted. Exactness means using the same words, same punctuation, the same spelling and the same capitalization. Complete accuracy is essential.

2. When a quotation is very long, or where a student wants to use only a few paragraphs or sentences within a larger passage, it is permissible to omit sections of an original passage. It should be used with extreme care so that the tone, meaning and intention of the original extract are not altered.

HOW TO QUOTE

The conventions adopted by different departments vary, but unless given specific directions in the contrary, there are a number of general procedures to follow in quoting.

Short Quote vs Long Quote

The basic form of quotation is initially determined by its length.

1. *Short quotation*

i. Incorporate the quotation into a sentence or paragraph framework, without disrupting the flow of the text.

ii. Use double quotation marks at the beginning and the end of the quotation.

iii. Use the same spacing as the rest of the text.

2. ***Long quotation***

    i. Use no quotation marks at the beginning and end of the quotation

   ii. Use single line spacing

  iii. Introduce the quotation appropriately

  iv. Indent the quotation three spaces from the left margin.

23

USING PUNCTUATIONS

Punctuation refers to the marks in writing used to separate sentences and their elements, and to clarify meaning. The following are the principal punctuations—full stop, comma, semicolon, colon, note of interrogation and note of exclamation.

FULL STOP OR PERIOD

Full stop represents the greatest pause and separation. It is used to mark the end of a sentence and to mark abbreviations and initials such as M.A., M.P., etc. They are used

- ◎ after a sentence,
- ◎ after an abbreviation of a state (Kerala—Kera.), of a month (Sept.), and of a day (Mon.),
- ◎ with an abbreviation in which omission of a period might cause ambiguity (confusion) like Fig., ed., eds.,
- ◎ with an abbreviation for a Latin term, e.g. i.e., and
- ◎ when non-technical words are abbreviated, e.g. Jr., Sr.

Do not use periods in the following situations:

- ◎ with a capital letter abbreviation, e.g. USA (not U.S.A.)
- ◎ with a symbol of a chemical, biological or medical expression (e.g. ATCH, ECG, EEG, DNA, RNA, RBC)
- ◎ with lower case abbreviations, e.g. diameter (diam), molecule (mol), mol wt., expt.
- ◎ with an abbreviation of unit measure, e.g. mg, g, cm, kg.

- ◎ after an item in the list of
  - i. Adrenalin
  - ii. Levamisole
  - iii. Acetylcholine
- ◎ after the end of a title or title in tables.

COMMA

The comma provides separation or a brief pause within a sentence and is helpful in grouping words, phrases and clauses for clarity and ease of reading. Do not separate a subject and its adverb or a verb and its object, except by phrases between commas. A comma is generally not placed before the word preceded by "and".

Use a comma

- ◎ to separate a series of words in the same construction, e.g. England, France, Italy formed an alliance;
- ◎ to separate each pair of words connected by "and", e.g. high and low, rich and poor, wise and foolish must all die;
- ◎ to mark a noun or phrase in apposition, e.g. Milton, the great English poet, was blind;
- ◎ before and after words, phrases (on the contrary, on the other hand, in fact, after all, in the first place) or clauses, let into the body of a sentence, e.g. He did not, however, gain his object;
- ◎ to separate adjacent sets of figures, e.g. "In 1935, 100 people died of cholera" and
- ◎ to group numbers in thousands, e.g. 1,000; 18,000, 1,00,000.

SEMICOLON

The semicolon represents a pause of greater importance than that shown by the comma. It separates segments of a sentence that are "farther apart" in position, or meaning, but which are nevertheless related. If the ideas were "closer together", a comma would have been used. It is also used to separate two clauses that may stand on their own but which are too closely related for a colon or full stop to intervene between them.

Use semicolon to separate

◎ the clauses of a compound sentence, when they contain a comma, He was a brave, large-hearted man; and we all honoured him;

◎ to separate a series of loosely related clauses as God gave her peace; her land reposed and

◎ to separate elements in between authors, e.g. A.K. Anderson; Raman, G.K. and Krishnan, N.

COLON

Colon marks a still more complete pause than that expressed by the semicolon. The colon is used before one or more examples of a concept, and whenever items are to be listed in a visually separate fashion. The sentence that introduced the itemized list you are now reading ended in a colon. It may also be used to separate two fairly—but not totally—independent clauses in a sentence.

Use a colon

◎ to introduce a long quotation;

◎ to introduce a list or enumeration;

◎ to separate parts of ratios;

◎ in literature citations to separate volume and page numbers and

◎ between sentences grammatically independent but closely connected in sense.

QUESTION MARK (NOTE OF INTERROGATION)

This is used after a direct question. Use question mark at the end of a direct question, even if the questions are presented in declarative form. Do not use a question mark after an indirect question. Place the question mark inside the quotation mark, if the question mark is part of the quotation, or outside if the mark is not part of the quotation.

NOTE OF EXCLAMATION

This is used after interjections and after phrases and sentences expressing sudden emotion or wish, e.g. What a terrible fire this is!

DASHES

Use dashes sparingly

- ◎   to indicate an abrupt break or shift in thought;
- ◎   to isolate parenthetical matter;
- ◎   within brackets and within parenthesis for a third level of interpolation;
- ◎   to resume a scattered subject, (e.g. friends, companions, relations—all deserted him.) and
- ◎   to indicate a range (e.g. 9–14).

PARENTHESES

This is used to separate a word(s) from the main part of the sentence, phrase or clause which does not grammatically belong to it.

Use parenthesis

- ◎   to set off a comment or explanation that is structurally independent of the sentence;
- ◎   to group mathematical expressions and
- ◎   to label enumerations included within a paragraph, e.g.

  i) ....   ii) ....   iii) ....

Avoid use of double parentheses. Use single parenthesis to set off enumerated paragraphs, e.g. 1), a), i).

BRACKETS

Use brackets

1. to set off words or other matter you have inserted in a quotation;
2. around parenthetical remarks inserted within other parenthetical remarks and
3. to set off bibliographic details not shown in the original.

APOSTROPHE

This is used a) to show the omission of a letter or letters as in Don't, e'er, I've and b) in the genitive case of a noun.

Use an apostrophe

◎ and 's' ('s) to form the possessive of a singular noun;

◎ to indicate omission of letters in a contraction; but contractions such as can't, don't and won't are inappropriate in scientific writing and

◎ 's' to form the plural of a letter, figures and some words. "He uses too many but's".

Do not use an apostrophe in certain well-established geographic names (Sims Park) or in names of some organizations (Teachers Association).

QUOTATION MARKS

Inverted commas are used to enclose the exact words of a speaker or a quotation.

Use double quotation marks in the text around

◎ all direct quotations;

◎ titles of articles, parts of books, and series titles and

◎ new technical terms or old terms used in a new unusual sense.

If quotation occurs within a quotation, it is marked by single inverted commas. Use single quotation marks around a word, title or term within a quotation.

HYPHEN

A shorter line than the dash (-) is used to connect the parts of a compound word, e.g. passer-by.

Use a hyphen

◎ between the numerator and denominator of a fraction when spelled out (one-third);

◎ between the parts of some compound words;

◎ between numbers to indicate a range and

◎ to connect parts of a word divided at the end of a line.

Avoid use of hyphen

- between parts of a compound modifier, if the modifier follows the noun modified and
- between words of a well-established open compound noun that is used to modify a substantive.

RAISED PERIODS

- To indicate multiplication, e.g. $k \times g = k \cdot g$
- To associate basal pairs of nucleotides, e.g. $A \cdot T$, $G \cdot C$
- To indicate water of hydration in chemicals, e.g. $CuSO_4 \cdot H_2O$

24

PROOFREADING

The manuscript usually goes through a copyediting procedure during which spelling and grammatical errors are corrected. Do not read proof only for sense; look at every letter and punctuation mark. Sometimes a typesetter's error can change one word to another. For example, "causal relationship" can become "casual relationship", "ingenuous" (innocent) to "ingenious" (clever) and "alternatively" (choice between two things) to "alternately" (to perform by turns). See that no opening or closing quotation marks or brackets are missing. Check the preliminary pages, headings and numerical sequence of pages, tables and illustrations.

Use standard symbols, write your corrections clearly. Always make it absolutely clear how much is to be deleted.

Mark in text	Mark in margin	Instructions
Hela cells	*caps* or *l.c*	Capitalize
The Penicillin reaction	*lc*	Make lower case
A very good reaction		Delete
MacDonald reaction		Close up
...in the cells. The next	¶	Start a new para
...in the cells after which		Insert comma
In the cells however		Insert semicolon
Well known event		Insert hyphen

(Contd.)

Mark in text	Mark in margin	Instructions
...in the cells☉ Then	⅄(·)	Insert period
in ᵗʰᵉcells	#the#	Insert word
Proofrᵉa̷der	↻	Transpose
CO_2	*Subscript*	Subscript
^{32}P	*Superscript*	Superscript
The(bacterium)was	*rom*	Set in roman type
P. aeruginosa cells	*ital*	Set in italics type
Results	*b.f*	Set in boldface type
A very good reaction	*Stet*	Let it stand
Do no effect	*t/a*	Add a letter
⊐Introduction⊏	⊐⊏	Centre

COMMON UNITS AND THEIR SYMBOLS

Unit	Symbol
Hours	hr
Minutes	min
Seconds	sec
Molar	M
Litre	l
Millilitre	ml
Grams	g
Kilogram	kg
Milligram	mg
Microgram	µg
Metre	m
Centimetre	cm
Kilometre	km
Per cent	%

Versus	vs
Volt	V
Volume	v, vol
Watt	W

COMMON ABBREVIATIONS

Article	Art
Augmented	aug
Book	Bk
Bulletin	bull
Copyright	c
Confer, compare	cf
Chapter(s)	chap(s)
Column	col
Editors	eds
Editor	ed
Note, Footnote(s)	n, nn
Figure	fig
Enlarged	enl
Line, Lines	l, ll
Manuscript(s)	ms, mss
No date of publication	n.d
Number(s)	no(s)
Page(s)	p, pp
Paragraph	par
Part(s)	pt(s)
Revised	rev
Sections	sec
Translator, -lation	trans

LATIN WORDS AND PHRASES

anon (anonymous)	used when the author is not known
c or ca *(circa)*	approximately, about
e.g. *(Exempli gratia)*	for example
et al. *(Et alli)*	and others
et alibi	and else where
et seq. *(Et sequens)*	and the following
Ibid *(Ibidem)*	in the same place
	e.g. *Ind. Vet. J.* 8: 16–20 *Ibid* 96–101
Idem	same (used in the case of same authors or same page)
	e.g. Krishnan, K and A. Radha (1960) *Idem* (1975)
i.e. *(Id est)*	that is
infra	below
loc. cit *(loco citato)*	in the place cited
NB *(Nota bene)*	note well
n.p	no place of publication given
op. cit *(Opere citato)*	in work already cited
passim	throughout; here and there
p.; pp.	the page; pages following
q.v.*(quod vide)*	which see
supra	above
viz., *(vide licet)*	namely
vid or vide	see

25

PHOTOGRAPHY

Photography includes all processes associated with making and recording actions of solar energy on matter. Photographs are required by the scientists to make an effective mark on oral communication.

ADVANTAGES OF PHOTOGRAPHS

1. Self explanatory—A photograph is equivalent to 2000 words.
2. Photos have universal language and can be appreciated by persons of different types and abilities.
3. By multiple copies, photos can reach many hands.
4. The birth of sciences is greatly facilitated because proof can be provided for observations.
5. It helps to convince others.
6. Data can be stored in microfilms.
7. Photos are simplified interpretations.

Photographs are a way of marking permanent, those passing images that catch your eyes. The camera is an essential equipment for photography. Camera and eyes have certain things in common:

1. both need light;
2. a lens organizes the light rays;
3. an image forms in both, i.e., on the retina in the eye and the emulsion-treated side of the film in the camera and
4. a message is received and recorded in the camera whereas the message is recorded and interpreted by the eyes.

TYPES OF CAMERAS

The prospective photographer is faced with the choice of three basic types of camera, the fixed lens range finder camera, the twin-lens reflex camera and the single-lens reflex camera.

Fixed Lens Range Finder Camera

Also known as viewfinder camera, this has separate lens for viewing an object and taking photographs. At close range, the image area seen through the viewfinder is not the same as that covered by the lens-parallax error.

Disadvantages

1. parallax error due to different positions of the eyes and camera lens;
2. minimal control over focusing and object sizes and
3. not possible to take close-ups; rarely focuses to less than 60 cm. The camera is cheap and easily handled.

Single-Lens Reflex Camera (SLR)

This is the most popular type of camera. A single lens is used for viewing and taking the picture. This eliminates parallax error. There is lens control over the setting and shutter speed. Depth of the field can be checked. In SLR, the image is viewed through the photographic objective lens so that the photographer sees exactly what will be recorded on the film. This has interchangeable lens, allowing introduction of extension tubes or close-up zoom lenses to increase image magnification.

Twin-Lens Reflex Camera

Two lenses are there, one exclusively for viewing and another for recording the image. The viewing lens allows more light to pass through the taking lens. The film size is much bigger, hence bigger negatives are obtained. Therefore, better pictures and any enlargement can be taken. Although excellent for photography at normal distances, it is more difficult to use at close range since matched pairs of close-up lenses have to be used.

Polaroid Cameras

This consists of a camera plus a photographic laboratory. Print is available immediately after taking a snap. There will not be sufficient clarity in the photographs.

LENSES

The lens system causes refraction of rays. The positive lens bends the light rays and causes it to converge. Negative lens makes the rays diverge. Cameras usually have compound lenses.

Wide-Angle Lens

The working angle of view of the human eye is between 45° and 50°. A lens that accepts an angle of view between 45° and 50° is called as normal lens or a standard lens. The standard lens with which SLRs are normally supplied usually produces an image through the viewfinder approximately to life size with a 35 mm focal length, which is capable of focusing from infinity to a distance of less than 60 cm from the subject. A wide-angle lens takes more image than the normal eye. The lens includes more of the subject than that of the standard lens. So the photographer can stay close to its subject and can photograph effectively. The focal length of a wide-angle lens is substantially shorter than the focal length of the normal lens for the image size produced by the camera. In the case of a 35 mm film, the snap size is 36×24 mm.

Narrow-Angle Lens

The lens angle is less than the eye of an average person. Its focal length is the same as that of the long side of the film foreman, i.e., 108×96 cm. This enables the photographer to reach out to the subject from long distance. This is used in wildlife photography.

Telephoto Lens

The focal length here exceeds 300 million. Photographs of very distant objects can be taken.

Zoom Lens

In this lens system, there is variability of focal length within a single lens. Long shots of close-up can be taken with the same lens. But, it is very expensive.

Fish Eye Lens

It covers objects lying over an angle of 180°. Maximum area is covered by this lens.

Close-up photography covers the image-subject magnification range from 1 : 10 to 1 : 1. The range beyond life-size reproduction is handled by a microscope (10 : 1)—photomicrography.

For getting close-up pictures, we can place supplementary lenses in front of existing camera lens or increase the distance between the lens and the film plane by inserting extension tubes or bellows.

APERTURE

This controls the actual amount of light that is made available. The 'f' number indicated in the camera are graduations to indicate how much light the aperture will allow through the lens. The aperture numbers are indicated as 2.5, 2.8, 3.5, 4.0, 4.5, 5.6, 6.7, 8.0, 9.5, 11.0, 13.0, 16.0, 19, 22, 27, 32.

$$F = \frac{\text{Focal length}}{\text{Working diameter}}$$

$F = 8$ means that the size of the opening is 1/8 of the focal length of the lens. Larger the 'f' number, lesser will be the light that passes.

SHUTTER SPEED

This controls the time of exposure of the light into the camera. It is given as 1/1000 to 1/10th of a second. In noon light, the aperture should be narrow, shutter speed should be 1/1000th. The shutter speeds are 30, 20, 15, 10, 1/4, 1/25, 1/45, 1/60, 1/90, 1/125, 1/180, 1/250, 1/350, 1/750, 1/1000, 1/1500, 1/2000th sec.

DEPTH OF FIELD

When your subject is in focus, there is a certain area in front of it and behind it, which will also be in focus. This range of sharpness is called "depth of field". The depth of field has the following characteristics:

1. the smaller the aperture, the wider the depth of field and vice versa;

2. the shorter the lens focal length is, the greater the depth of field, provided the aperture and shooting distance are the same;

3. the further the shooting distance, the greater the depth of field and

4. depth of field is generally greater in the background than the foreground.

FILMS

Generally, 35 mm films are used. The speed of the film relates to the swiftness with which the object falls in emulsion. Faster the film, less is the exposure time. If very fine details are required, use only slow-speed film. The fast films have large particle size for the silver emulsion in the film. Slower-speed films will have smaller sized particles. The speed is expressed with ASA, which is an index of the amount of light needed for correct exposure of the film. The lower the film speed, the greater the exposure needed, the higher its contrast and the finer its grain. Theoretically, therefore, the slower the film the better its resolving power. This ranges from 100 ASA to 5000 ASA (American Standards Association).

Generally, negative films are used for taking prints whereas positive films are used for preparing colour transparency slides. Colour negative film allows colour prints to be made directly, much as in black-and-white photography. Colour negatives are less sharp than black-and-white negatives. Colour reversal film produces a positive transparency. AGFA, KODAK, KONICA, FUGI, SAKURA are some of the trade names of the films available.

DEVELOPING

By this process, a latent image is converted to real image. It is a chemical process done in the dark room.

1. Dip the film in water for 1 min. wetting.

2. Dip in developing solution and keep rolling for 3–7 min. Developing solution contains metol (2.2 g), hydroquinone (17 g), sodium sulphate (anhydrous) (7.5 g), sodium carbonate 65 g, potassium bromide 2.8 g and distilled water to make 1L. For use, it is diluted with water in 1 : 3 proportions. Metol and hydroquinone react with silver halide; sodium carbonate accelerates the process of developing. Potassium bromide acts as a destainer. This prevents over-developing. Sodium sulphate is a preservative. Temperature should be 20°C.

3. Dip in acid alcohol (2% acetic acid for 2 min.).

4. Fixation is done in sodium thiosulphate solution (1–2%).

5. Wash in running tap water for 20 min., allow to dry. Negative is obtained in 35 min.

PRINTING

From the negatives obtained, we make prints. There exist many methods of printing.

1. In **contact printing,** the negative is directly placed on the printing paper without any enlargement. This is less expensive.

2. In **project printing,** negatives are enlarged to different sizes like 2B, ¼, ½ and full size.

3. In **paper printing**, different types of papers (soft and hard paper) are used for taking prints. Soft paper is used when high-contrast negatives are available. Hard paper is used for low-contrast negatives, particularly for photomicrography.

4. **Glossy printing** is done for gross photography and **matt printing** is done and for scientific photography.

26

WRITING
A RESEARCH PROPOSAL

The research proposal gives details about the research work to be undertaken by the individual, with specific information on the topic of research, methodologies to be adopted and results anticipated. The researcher should be clear about what his research is and how he is going to execute the work. He should indicate what he is anticipating from his research. Before starting to write the research proposal, one should get a good idea of the topic, which could be obtained by thorough literature search. Topics could be identified from either research papers published or by conversing with other research scientists. Once an idea has been obtained about the proposed research area, other information could be gathered and writing could be started. A research proposal is intended to convince others that the proposal is a worthwhile research project and that the presenter of the proposal has the competence and the work plan to complete it. Generally, a research proposal should contain all the key elements involved in the research process and include sufficient information for the readers to evaluate the proposed study. Regardless of the research area and the methodology chosen, all research proposals must address the following questions: What is planned to be accomplished, why it is done and how it is going to be done.

A research proposal needs to show how the current proposal fits into what is already known about the topic and what new contribution this work will make. The question that would be answered by the research should be specified and how this work would be carried out should be elaborated. The proposal should situate the work in the literature, it should show why this is an important question to answer in the field and the committee should be convinced that this approach would in fact result in an answer to the question.

The quality of a research proposal depends not only on the quality of the proposed project, but also on the quality of the proposal writing. A good research project may run the risk of rejection simply because the proposal is poorly written. Therefore, it pays if the writing is coherent, clear and compelling.

The research proposal may contain 2–3 pages of a preliminary proposal or it may contain 30–50 pages of detailed research proposal with different sections. In the preliminary proposal, in the first paragraph, the first sentence identifies the general topic area. The second sentence gives the research question and the third sentence establishes its significance. The next couple of paragraphs give the larger historical perspective on the topic. Essentially list the major importance of the topic and very briefly review the literature in the area with its major findings. The question to study may be restated showing how it fits into this larger picture. The next paragraph describes the methodology. It tells how the question would be approached. The final paragraph outlines the expected results, how it will be interpreted.

The regular standard research proposal contains the following sections:

- Introduction
- Literature review
- Methodology
- Expected results
- Budget
- References

INTRODUCTION

The topic should be first chosen for the research proposal. A good title will clue the reader into the topic, but it cannot tell the whole story. The topic should be concise and descriptive and should not be too broad or too narrow. The introduction provides a brief overview of the whole research proposal. It might be as short as a single page, but it should be very clearly written. The introduction typically begins with a general statement of the problem area, with a focus on a specific research problem, to be followed by the rational or justification for the proposed study. Few sentences on the general area of the topic are given, followed by more specific statements about the research topic. The questions that would be addressed by the work and the importance of the work should be given.

Once the topic is identified, the work to be undertaken, specific questions to be answered and the results anticipated should be prepared. The importance of the work is to be highlighted. The proposal should also give what new knowledge would be obtained from this study. Some questions that are to be considered are:

- Is the question an important one?
- Will the results change the current practice?
- Will the results lead to better applications, procedures or outcomes?

Research proposal should state the major purpose of the actual research showing the issues that need further investigation. The rationale for the research should be clearly indicated. The major issues and problems to be addressed by the research should be mentioned.

LITERATURE REVIEW

The main purpose of the literature review is to situate the present research proposal in the context of what is already known about the topic. It should provide the theoretical basis for the work, show what has been done in the area by others, and set the stage for the present work. Performing a literature review is also a good method of developing a question. Picking a topic of interest and examining what research has been performed previously may be an effective way to do this. Reviewing published literature may provoke an idea or question by suggesting what areas of a topic should be explored further.

Review of literature will indicate the gaps in knowledge that need to be filled in. One should indicate that the present proposal would fit in and would contribute for newer knowledge. The preliminary research work carried out by the proposal submitter may be given here. It would impress the committee that the proposal submitter has got the required research experience and knowledge on the subject matter.

Hypotheses should be formulated to guide the research work. The purpose of the hypothesis is to help guide scientific inquiries and often follows directly with the question being asked or theory being tested. It would help to provide direction to the research design and assist in the collection, analysis and interpretation of data. A hypothesis is a tentative prediction or explanation of relationship between two or more variables and this statement describes the results the researcher expects to obtain. Research studies should be designed to test the hypothesis, and the nature of the hypothesis will help determine the sample group for study, measuring instruments, study design, procedures and statistical techniques used for data analysis.

The proposal should then contain the study design to obtain answers to the questions being asked and the hypothesis being tested. It spells out strategies to develop information that are accurate, objective and meaningful, and explains the methods that will be used to collect and analyse data. This also should include the time frame necessary or allowed to conduct the study.

METHODOLOGY

The Methods section is important because it tells how the research is going to be carried out and it provides the work plan and describes the activities necessary for the completion of the project work. The methodology should include all the materials and equipment that would be used in the research, the methodology that would be adopted, the data that would be collected and the statistical methods that would be used for analysis of data. A methodology contains a number of methods that would be used including qualitative and quantitative studies, case study, experiments, sampling, etc. How the anticipated outcomes will be interpreted to answer the research question should be given.

An important part of the proposal is planning the research in all its stages up to completion. Research plan should specify what tasks will be completed at each stage—literature review, research framework, description of method, writing up of findings and conclusions and so on. These tasks should specify what writing tasks will be accomplished and when.

EXPECTED RESULTS

This section should give a good indication of what result is expected to be obtained out of the research. It should join the data analysis and possible outcomes to the theory and questions that have been raised earlier. It would be a good place to summarize the significance of the work. It is often useful from the very beginning of formulating the work to write one page for this section to focus the reasoning for the proposal.

BIBLIOGRAPHY

This section should list all the relevant references, which are included in the text. Complete references should be given with the author(s) name, year, title of publication, volume number and starting and ending pages.

27

WRITING A RESEARCH REPORT

Research reports are written to communicate the results of research, field work, and other activities. Often, a research report is the only concrete evidence of research, and the quality of the research may be judged directly by the quality of the writing and how well the findings are conveyed. Fortunately, research reports, which are similar to research articles, technical reports, lab reports, formal reports, or scientific papers (to name a few), have a fairly consistent format that will help you to organize your information clearly. Most research reports contain the same major sections, although the names of the sections vary widely, and sometimes it is appropriate to omit sections or add others. If you are submitting a research report for a class or to an organization, check for specific requirements and guidelines before beginning to write your research report.

The research report enables to demonstrate the writer's ability to conduct a literature search on a selected topic, critically review the literature, compare, contrast, and integrate the findings, identify areas for further research, and design a study to investigate at least one of those areas.

Most of the research work should have been completed before writing of the report is started. It is important to consider the audience before starting to write the research report so that the report will adequately communicate the research results and its significance to the readers. Report could be written in such a language that it is understood clearly by the audience. The readers will be motivated to read the research report for a variety of reasons, in general they will read to:

- learn about research related to their particular research interests,
- keep abreast of research in the discipline in general,
- keep current with research related to their teaching interests and
- keep informed about the scientific literature in related disciplines.

SECTIONS OF A RESEARCH REPORT

Reports are generally divided into sections. Each section has a specific purpose, and often there are specific guidelines for formatting each section. Generally, a report will include the following sections:

- Title Page
- Abstract
- Table of Contents
- Introduction
- Body
- Recommendations
- References
- Appendices

Title Page

The title page of the research report normally contains four main pieces of information:

- the report title;
- the name of the person, company, or organization for whom the report has been prepared;
- the name of the author and the company or university which originated the report; and
- the date the report was completed.

Title of the Report

It is a good idea to develop a "working title" for the project as the initial draft of the report is prepared. Be sure that the title is accurate; it needs to reflect the major emphasis of the paper and prepares readers for the information presented. Also, the title developed should be interesting to readers and should make them want to read the rest of the report.

There are four common approaches that writers often take in writing their titles.

◎ Include the name of the problem, hypothesis, or theory that was tested or is discussed.

◎ Include the name of the phenomenon or subject investigated.

◎ Name the method used to investigate a phenomenon or method developed for application.

◎ Provide a brief description of the results obtained.

Obvious words and phrases such as "A study on ..." and "An investigation of ..." should be omitted whenever possible. These make the title unnecessarily wordy.

Abstract

"An abstract is an accurate representation of the contents of a document in an abbreviated form". An abstract can be the most difficult part of the research report to write because it is in the abstract that the subject matter would be introduced, the work done would be given, and selected results are presented, all in one short (about 250 words) paragraph. As a result, an abstract is usually written last. An abstract serves an important function in a research report; it communicates the scope of the paper and the topics discussed to the reader, and, in doing so, it facilitates research. Abstracts help scientists to locate materials that are relevant to their research from among published papers, and many times scientists will read only a paper's abstract in order to determine whether the paper will be relevant to them.

The most common type of abstract is the informative abstract. An informative abstract summarizes the key information from every major section in the body of the report, and provides the key facts and conclusions from the body of the report. A good way to develop an informative abstract is to devote a sentence or two to each of the major parts of the report. Key numerical facts should be included to make the informative abstract brief.

Table of Contents

Most reports will contain a Table of Contents that lists the report's contents and demonstrates how the report has been organized. Each major section is listed in the Table of Contents. Additional descriptive headings are used sometimes throughout the report and for the Table of Contents. Using descriptive headings can help readers to see how the report is organized if the section headings are not clear enough.

Introduction

The introduction prepares readers for the discussion that follows by introducing the purpose, scope, and background of the research. The audience for the report largely determines the length of the introduction and the amount of detail included in it. Enough details should be included so that someone knowledgeable in the field can understand the subject and research.

It is very important to mention the purpose of the research and report it in the introduction. The following questions will help to think about the purpose of the research and reason for writing a report:

- What did the research discover or prove?
- What kind of problem was investigated?
- Why was this work on this problem carried out? If the problem was assigned, why did the instructor assign this particular problem; what does one learn from working on it?
- Why was this report written?
- What should the reader know or understand when they have finished reading the report?

Scope refers to the ground covered by the report and will outline the method of investigation used in the project. Considering the scope of the project in the introduction will help readers to understand the parameters of the research and the report.

The following questions will help to think about the scope of both the research and the report:

- How was the research problem studied?
- Why was the work tackled in the manner presented?

- ◎ Were there other obvious approaches that could have been taken to this problem?

- ◎ What were the limitations that were faced that prevented the researcher trying other approaches?

- ◎ What factors contributed to the way the researcher worked on this problem? What factor was most important in deciding how to approach the problem?

The Body of the Report

The body is usually the longest part of the research report, and it includes all of the evidence that readers need to have in order to understand the subject. This evidence includes details, data, results of tests, facts, and conclusions. In general, the body of the research report will include three distinct sections:

- ◎ a section on theories, models and the hypothesis,

- ◎ a section in which the materials and methods used in research are discussed and

- ◎ a section in which the results of research are presented and interpreted.

Theories, Models, and Hypotheses

This section may be included to discuss the theories and models upon which the research project is based. This section can be very important, especially for research articles, formal reports, or scientific papers, but sometimes it will not be required for lab reports and other homework assignments. Regardless of whether this section on theories and models are included, the research will include models and theories that other researchers have developed. In this section of the paper the data found in the study will not be discussed; instead, the theoretical basis for the project will be introduced.

In this section, the hypothesis and the theories and models that are used to develop it should be explained; the competing hypotheses, theories, and models, including their strengths and weaknesses should be defined and explained and the specific points where they agree or disagree should be compared and contrasted.

Materials and Methods

The materials and methods section is similar to an instruction manual. It should describe the apparatus and the procedure that are used in the experiments. This section should be clearly and specifically written; another researcher should be able to exactly duplicate the research performed by following the procedures outlined in this section.

All materials and methods sections should address the following questions:

- How was the experiment designed?
- On what subjects or materials was the experiment performed?
- How were the subjects/materials prepared?
- What machinery and equipment were used in the experiment?
- What was sequence of events that were followed, the subjects/ materials that were handled and the data recorded?

Any special or nonstandard equipment used in the study should be described. Illustration of the equipment may be provided and the method used for recording the data may be given. The species, genus, strain, and breeding origins of any animal used should be included. The amount and purity of any chemicals used should be provided. The details of the exact and original experimental procedure should be described.

Methodology

It must be written clearly so that it would be easy for another researcher to duplicate your research if they wished to. It is usually written in a passive voice (e.g. the participants were asked to fill in the questionnaire attached in Appendix 1) rather than the active (e.g. I asked the participants to fill in the questionnaire attached in Appendix 1). It is written in past tense. Ensure that any material that has been used from another source should be clearly cited as reference. Any diagram, chart or graph used should be labelled and numbered.

Results

In this section of the report, detailed results of the research should be given along with experimental data, observations, and outcome.

All preceding sections of the report (Introduction, Materials and Methods, etc.) lead to the Results section of the report and all subsequent sections will consider what the results mean (discussion, conclusion, recommendations, etc.). The most common way to organize information in a research report is to list it chronologically. This method of organization allows you to present information in the sequence that events occurred. Another good way to help organize information so that readers will understand what is most important is to give it in a figure or table. Complex data should be presented in the form of tables and figures.

Figures and Tables

Most scientific reports will use some type of figures and/or tables to convey information to readers. Figures visually represent data and include graphs, charts, photographs, and illustrations. Tables organize data into groups. Figures and tables should help to simplify information, so one should consider using them when words are not able to convey information as efficiently as a visual aid would be able to.

It is important to choose the correct way to represent the data depending on the type of audience for the report. Tables or lists are simple ways to organize the precise data points themselves in one-on-one relationships. A graph is best at showing the trend or relationship between two dimensions, or the distribution of data points in a certain dimension (i.e., time, space, across studies, statistically). A pie chart is best at showing the relative areas, volumes, or amounts into which a whole area (100%) has been divided. Flowcharts show the organization or relationships between discrete parts of a system. For that reason they are often used in computer programming. Photographs are best at presenting overall shapes, shades, and relative positioning, or when a "real-life" picture is necessary, as in the picture of a medical condition or an electron micrograph of a particular microscopic structure. Illustrations are best when they are simple, unshaded line drawings. They suit most purposes for representing real objects or the relationship of parts in a larger object.

Additionally, all tables and figures should be self-contained—they should make complete sense on their own without reference to the text; be cited in the text—it will be very confusing to the audience to suddenly come upon a table or figure that is not introduced somewhere in the text. They will not have a context for understanding its relevance to the report; numbers such as Table 1 or Figure 10 should be included which will help to distinguish multiple tables and figures from each other and include a concise title—it is a good idea to make the most important feature of the data the title of the Figure and Table.

Discussion

This section of the report is important because it demonstrates the meaning of the research. In this section results are interpreted and the current research results are discussed in relation to the findings of earlier research. This section of the research report begins with a discussion of the data and then the data are analysed. How the data addresses the research problem or hypothesis should be outlined in the Introduction and subsequently what can be inferred from the data as they relate to other research and scientific concepts should be discussed.

It is also very important for the researcher to identify the nature and extent of any limitations of the research in this section of the report, especially if the results are inadequate, negative, or not consistent with earlier studies or with the hypothesis. Attempts should be made to defend the research or minimize the seriousness of the limitation in the interpretation.

Conclusions

The conclusion is important because it is here that the significance and meaning of the research is conveyed to the reader by concisely summarizing the findings and generalizing their importance. It is also a place to raise questions that remain unanswered and to discuss ambiguous data. The Conclusion follows naturally from the interpretation of data, so, in some cases, a separate section "Conclusions," will not be required, but the discussion can be ended with the conclusions. The most important thing to remember in writing the conclusion is to state the conclusions clearly. The unanswered questions may be raised and the ambiguous data could be discussed. Raising questions or discussing ambiguous data does not mean that the work is incomplete or faulty; rather, it connects the research to the larger work of science and parallels the introduction in which questions are raised.

Recommendations

This section appears in a report when the results and conclusions indicate that further work needs to be done. If a Recommendations section is included, there is another opportunity given to demonstrate how the research fits within the larger perspective of science, and the section can serve as a starting point for future dialogue on the subject. It demonstrates that the importance and implications of the research have been fully understood.

References

It is important to include a References section at the end of a report in which other sources of information have been used. Reference sections are important because they allow other researchers to build on or to duplicate the research. Without references, readers will not be able to tell whether the information provided is credible, and they will not be able to find it for themselves. Reference sections also allow one to refer to other researchers' work without reviewing that work in detail. Reference styles vary greatly from one instructor to another, and one journal to another. The reference should be formatted according to the guidelines provided by the journal or teacher to whom the report is submitted.

The information to be included in the reference list are the following.

◎ Author's name or authors' names

◎ Title of the document

◎ Identification information

  ◎ **Books**   City, state, or country of publication, publisher's name, and year of publication. Editor's name, chapter title and author, and page numbers of chapter, if applicable.

  ◎ **Journal articles or technical papers**   Journal's name, volume and issue number, date of issue, page numbers of referenced articles.

- **Reports** Report number, name and location of issuing organization, date of issue.

- **Correspondence** Name and location of issuing organization; name and location of receiving organization, letter's date.

- **Conversation, conference presentation, or speech** Name and location of speaker's organization; name, identification, and location of listener; date.

Appendices

Some information may be relevant to the subject of research, but needs to be kept separate from the main body of the report to avoid interrupting the line of development of the report. Such information should be placed in an Appendix. Anything can be placed in an appendix as long as it is relevant and as long as reference is made to it in the body of your report. An appendix should include only one set of data, but additional appendices are acceptable if several sets of data are needed that do not belong in the same appendix. Each appendix should be labelled with a letter, A, B, C, and so on.

28

WRITING
A RESEARCH GRANT PROPOSAL

The research grant proposal must be written in sufficient detail to allow reviewers to understand:

- what the project hopes to accomplish;
- if the project personnel have the necessary expertise to accomplish the goals and objectives;
- the potential of the project to improve undergraduate education;
- the national impact and cost-effectiveness of the project and
- the evaluation and dissemination plans.

A meticulous review of the guidelines will help to tailor the project proposal to the funding source's specific interests and will increase the likelihood that it will be funded. Proposals are most often funded based upon how well the project and the funding source's interests "match". A careful review of guidelines and a little background research will demonstrate the relevance of the project to the funding source's interests and objectives.

It should be remembered that those who review proposals are often reading a number of proposals in a short amount of time. Therefore, the proposal should be written with clarity. As with academic papers, enlisting the help of colleagues to read drafts can prove invaluable. Colleagues from within the same field are, obviously, most appropriate to review content, but colleagues from outside discipline are also welcome to review, because they may be less familiar with the subject area, and hence, they may be the best persons to judge if the proposal has been written in the most straightforward terms to convey the ideas.

PROBLEM STATEMENT

What is the problem or gap that this project responds to? They arise out of a situation in which there is a need for the project, a problem that must be solved. The problem situation should be discussed. One way of doing this is to think in terms of a lack, need or insufficiency in the situation. But remember—one must convince the audience that the problem is important for the audience and therefore deserves to be addressed.

OBJECTIVES

What will be the results of the project? The audience needs to know what outcome they can expect. The goals of the project are to be described in very specific terms. How the objectives will address the problems should be clearly shown. Often, formulating a hypothesis of the research problem would lead to listing of objectives.

GETTING STARTED

A good proposal begins with a clear idea of the goals and objectives of the project—for, creating a course or curriculum, for improving a laboratory by teaching new concepts directly, for teaching new material to undergraduate faculty, and so on. In addition, a good project begins with a sense of why it will be a significant improvement over current practice. Envision what improvements this project will make, and then develop the activities and course(s) accordingly.

METHOD

What procedures will be used to achieve the goals of the project? In empirical research projects, this is where the actual research being proposed could be described.

It is important that the Method section convinces the audience that

1. there is an effective means of achieving the stated objectives and

2. the researcher has a full understanding of how to manage the project.

After the goals and associated activities are well-defined, the resources (e.g. people, time, equipment and technical support) that would be necessary are to be considered. A better proposal is likely to result if the goals and activities are clear before resources are considered. Projects should explore teaching and learning methods that use equipment, scientific knowledge, or teaching techniques in effective ways; perhaps by adapting techniques to a new context or by teaching in a novel or attractive way.

RESOURCES

The amount of time and/or money and facilities required to complete the project should be assessed. This is where one makes the case for the resources that are needed from the funding agency. A list of the resources and how they would be used should be given. In complex projects, a detailed budget should be given. Mention what work has been done in preparation for the project. Evidence of preliminary work demonstrates planning and commitment to the project and often indicates the project's potential for success. When the proposal requests significant funds for equipment, it is helpful to consider alternatives and to explain why the instruments chosen are particularly suitable for the project.

Contents of a Proposal should include introduction, which gives the brief outline of the research to be undertaken with the hypotheses and objectives. Credentials of the researchers may also be given here. Then, the problem statement and need for research have to be given. A brief review of research done already should also be given with the gaps, that need further research work. The importance of the project is to be highlighted with the feasibility of solving the problem.

Research grant proposal is written with the following sections:

◎ Objectives

◎ Methods

◎ Evaluation

◎ Budget

◎ Abstract

GATHERING BACKGROUND INFORMATION

When writing a proposal, previously awarded projects or work supported in other ways should be studied. The relationship of the proposed project to the work of others should be described. In addition, the proposal must give appropriate attention to the existing relevant knowledge base, including awareness of current literature.

LOOKING AT THE PROGRAM ANNOUNCEMENT

The grant proposal that best fits with the programme announcement should be identified. The Program announcement guidelines should be carefully studied. Each program's section of that announcement specifies requirements for that proposal and information that is used to review the proposal. The Program announcement clearly spells out requirements, including format requirements. All parts of the proposal should conform to the requirements, i.e., target dates, font size, page limits, program objectives, budget limits, matching funds, etc. The proposal should be concise and not exceed any text restrictions.

WRITING THE PROPOSAL

A good proposal is always readable, well-organized, grammatically correct, and understandable. Be explicit in the narrative about how the program will make an improvement. This narrative must contain specifics including details of experiments and/or applications, both to show that planning has been done and to help reviewers understand why the particular application proposed is better than other ideas.

The narrative should be specific about the proposed activities. Reviewers want details of the project's organization, the course content, laboratory and other inquiry-based experiments, and participant activities, both to show that groundwork has been laid and to help them understand why the particular ideas that are proposed are better than others. Careful writing allows one to describe, in the limited space available, enough about the project to give the reviewers a clear idea of exactly what is planned to be done and why the plan is a good one.

The broad knowledge of the researcher in relevance to the grant proposal should be given. It is helpful to reviewers to see that an adequate timeframe has been provided for successful completion of the project. This will show that adequate planning has been done and that the program's implementation is realistic.

BUDGET INFORMATION

The budget request should be realistic for the project and reflect the goals of the project. It must also be consistent with the requirements of the funding agency. It should request sufficient resources needed to carry out the project, but it should not be excessively high. Budget information should be complete and unambiguous. The Program announcement should be consulted for eligible and ineligible items. Most reviewers and all Program Directors look carefully at the proposed budgets to find evidence of careful reflection and realistic project planning. Cost of the project must be realistic. Many budget requests are out-of-line with others submitted to the program. The Program announcement should be examined carefully for average size of awards and the award range. Budgets are often negotiated as a proposal is being considered; but a clear, realistic budget request strengthens a proposal. The time schedule for completing the project on time should be provided. This indicates a way of judging whether or not the project can be finished by the end of the research period. It also gives you a set of expectations to gauge the progress of the project.

CREDENTIALS OF THE PRINCIPAL INVESTIGATOR AND OTHER STAFF

When writing up the credentials of the faculty for the grant proposal, each biographical sketch should be written with the proposal in mind and should display the unique background of the principal investigator (PI) and other investigators, which will be valuable in working on the proposed project. In the format and length of biographical sketches the program guidelines should be carefully followed. It is important to address the experience and capabilities that the person or the project team will bring to the project. If it is one person, then the qualifications of that individual should be elaborated so that his qualifications fit the project. If it is a team effort, the qualifications of each member should be presented, especially showing how individual members have special skills or knowledge that can contribute to the project's success.

COST-BENEFIT ANALYSIS

Do the benefits of the project outweigh the costs? The proposal could be concluded with a brief statement of benefits versus costs. The financial data may be given that could persuade the audience that the expected benefits outweigh the costs. Cost-benefit analysis would make a final point about the value of the research grant proposal.

APPENDIX

PANEL DISCUSSION

It is a discussion in which few persons (the panel) carry on conversation in front of the audience. At the end of the discussion, the audience also participates. The audience put important questions and the experts answer them and clarify the points. The word "panel" means a group of experts. The discussion held among these expert members in front of an audience could be called panel discussion. The panel would change from subject to subject, but there would be an anchor person, who would introduce the panelists, receive questions and distribute them to the panelists upon their specialization.

A panel consists of a small group of six or eight persons, who carry on a guided and informal discussion before an audience as if the panel were meeting alone. The proceedings of the panel is volunteering of facts, asking questions, stating opinions—all expressed with geniality, with respect for the contributions of other members, without speech making, and without making invidious personal references. This primary function should occupy approximately two-thirds of the allotted time, say forty minutes of an hour's meeting. The secondary function of the panel is to answer questions from the audience.

Purpose

The purpose of panel discussion is to get important facts and different viewpoints out into open and to reproduce the features of a small discussion group for the benefit of a larger group. This discussion stimulates the thought process of the audience and lays a basis for wide participation later. In a panel discussion, people talk about a specific topic. A moderator guides the discussion by asking questions and making sure that everyone gets to speak. The purpose of a panel discussion, may simply be to provide information, or it may be to create a forum, a place for people to express their opinions and ideas.

Objectives

- To provide information and few facts
- To analyse a problem from different angles
- To identify the values
- To organize for mental recreation

There are two types of panel discussions—public panel discussion and educational panel discussion.

Public panel discussions are organized to discuss the problems of common man. Its objectives are:

- To provide factual information regarding current problems
- To determine social values
- To receive opinions of common men in topics of interest, e.g. annual budget, educated unemployment, increase in price of things.

Educational panel discussions are used in educational institutions to provide factual and conceptual knowledge and clarification of certain theories and principles.

Sometimes, these are organized to find out the solutions to certain problems.

Its objectives are:

- To provide factual information and conceptual knowledge
- To create awareness of theories and principles
- To provide solution to certain problems

Members in a Panel Discussion

Members in a panel discussion include the moderator or leader, panelists and audience.

Moderator or leader   They plan how, where and when the panel discussion will be organized. They prepare the schedule for panel discussion and sometimes the rehearsal also. The leaders must in addition take special care to select panel members who can think and speak effectively. They must also be sure that they prepare themselves to discuss the subject. During the discussion, the leaders should keep themselves more in the background as chairman of the panel. When the subject is thrown open to the house, it is the leader's job to recognize appropriate questions and to reject those not bearing on the subject or involving personalities. Some questions they may answer themselves, but usually they should repeat the question and call upon one of the panelists to answer it. By preliminary announcement, the leader may also tell the audience that they may direct questions at particular members of the panel if they choose. Leaders keep the discussion on the theme and encourage interaction among members, summarize and highlight the points and should have mastery on the theme or problem of discussion.

The moderator should

◎   introduce panel members and ask questions

◎   give everyone a chance to speak

◎   take questions from the audience at the end of the discussion

At the end of the discussion, the moderator summarizes the discussions, presents his/her point of view and finally expresses thanks to panelists and audience.

Panelists   There are 4–10 panelists in the discussion. Members sit in a semicircle in front of the audience. The moderator sits in the middle of the panelists. All panelists must have mastery of the subject matter. Panelists should

◎   stay tuned to the topic of discussion at all times

◎   answer each question when asked

◎   get involved in the discussions

Audience   Audience is allowed to ask questions and seek clarification. They can present their point of view and their experiences regarding the theme. The panelists answer. In some situations, moderator also tries to answer the questions.

How is a Panel Discussion Set Up?

◎   Experts in the field are invited as panelists.

◎   There is an invited audience of trainers and trainees or teachers and students.

◎ Questions on the proposed topic are collected in advance from the audience and handed over to the panelists.

◎ Questions are generally classified according to the sub-topic/ aspect/ dimension of the proposed theme. The panelists come prepared to the panel discussion.

How is a Panel Discussion Conducted?

◎ The moderator introduces the theme of the discussion to the participants. The moderator also initiates the discussion on the issue under consideration.

◎ The questions are then addressed to the panelists in a pre-determined order.

◎ The panelists are called (in a pre-determined order) to express their views.

◎ Later, panel members may react to, respond to, or complement the views expressed by co-panelists.

◎ At the end of the session, the moderator integrates and synthesizes the different points of view and presents them to the audience.

◎ If there is time, the moderator can call for further questions.

◎ Finally, the moderator sums up the discussion and highlights the key points.

Characteristics of Panel Discussion

◎ Used at the university and college level to organize teaching at reflective level.

◎ Develops the ability of problem-solving.

◎ Helps to understand the nature of the problem or theme of discussion.

◎ Develops ability of presentation of theme, giving their point of view logically.

◎ Develops right type of attitude and ability to tolerate anti-ideas of others.

◎ Helps in creative thinking.

◎ Develops manners of putting questions and answering them.

Advantage of Panel Discussion

◎ Encourages social learning

◎ Higher cognitive and effective objectives are achieved

◎ Used to develop ability of problem-solving and logical thinking

◎ Develops capacity to respect others ideas and feelings and ability to tolerate

◎ Facilitates clarification on knotty issues

◎ Highlights the multi-dimensionality of the issue under discussion

◎ Develops presentation skills

◎ Teaches students to think of the issues under consideration and ask relevant questions

Limitations of Panel Discussion

◎ There are chances to deviate from the theme of discussion.

◎ Some members may dominate.

◎ There is possibility to split the group into two subgroups ("for" and "against").

◎ If panelists belong to different groups, it may not create appropriate learning situation.

How to Organize Effective Panel Discussions

◎ There should be a rehearsal before the actual panel discussion.

◎ The moderator should be a matured person and should have the full understanding of theme/problem and should have full control over the situation.

◎ The seating arrangement should be such that each person is able to see and observe all the others without difficulty.

◎ Moderator should encourage constructive discussion among panelists and audience.

GROUP DISCUSSION

A group discussion (GD) can be defined as a formal discussion involving 8 to 10 participants in a group. A GD is an activity where groups of 8–10 candidates are formed into a leaderless group, and are made to analyse and discuss one of the following within a given time limit, which may vary between twenty minutes and forty-five minutes.

1. A specific situation
2. Case study for which they are asked to come out with a solution for a problem
3. Any topic

The seating arrangement for the group discussion should be such that all can face each other, i.e. in circles or squares. All should remain seated during discussion. Everyone should know each other. Blackboard and chalk should be ready for use if necessary. Discussions should be started and closed on time. Everyone should take part in the discussion. Preparations should be made for discussions beforehand. Questions to be asked in the discussion should be prepared in advance. Discussion should be conducted in a more friendly way.

Reasons for Having a GD

◎ It helps you to understand a subject more deeply.
◎ It improves your ability to think critically.
◎ It helps in solving a particular problem.
◎ It helps the group to make a particular decision.
◎ It gives you the chance to hear other students' ideas.
◎ It improves your listening skills.
◎ It increases your confidence in speaking.
◎ It can change your attitudes.

Do's for GD

◎ Speak pleasantly and politely to the group.
◎ Respect the contribution of every speaker.
◎ Remember that a discussion is not an argument. Learn to disagree politely.
◎ Think about your contribution before you speak. How best can you answer the question/contribute to the topic?
◎ Try to stick to the discussion topic. Don't introduce irrelevant information.

◎   Be aware of your body language when you are speaking.

◎   Agree with and acknowledge what you find interesting.

Don'ts for GD

◎   Losing your temper—a discussion is not an argument.

◎   Shouting—use a moderate tone and medium pitch.

◎   Using too many gestures when you speak—gestures like finger pointing and table thumping can appear aggressive.

◎   Dominating the discussion—Confident speakers should allow quieter students a chance to contribute.

◎   Drawing too much on personal experience or anecdote—although some tutors encourage students to reflect on their own experience, remember not to generalize too much.

◎   Interrupting—wait for a speaker to finish what they are saying before you speak.

A GD involves

1.   communication skills
2.   knowledge and ideas regarding a given subject
3.   capability to coordinate and lead
4.   exchange of thoughts
5.   addressing the group as a whole
6.   thorough preparations

The following points to be kept in mind by the participants of the GD:

1.   In one discipline, at least one person must be thorough in the subject

2.   Participants should speak freely—feel self-confidence in mind, your idea is also equally important, so feel free to express your ideas.

3.   Listen thoroughly to others, give respect, invite their opinion, and adopt patience.

4.   Jot down the points in a paper on which clarifications are required.

5. Discussions are carried out more informally—keep to your seat.

6. Do not monopolize the discussion; do not take more than a minute or two.

7. Make your point in a very few words.

8. Don't allow the discussion to get away from the point of discussion. You should fully concentrate and always be along with others. Do not keep aloof.

9. Indulge in friendly disagreement. Don't lose your temper. Check your language.

10. Strike while the idea is important. Don't wait for the leader to call you to express your viewpoints.

11. Come to the discussion with well-prepared questions. Since the topic is already known, read, concentrate, think and prepare yourself.

12. During the discussion, evidences like newspaper cuttings, etc., could be brought for emphasizing a point.

Points to Remember

◎ Knowledge is strength. A candidate with good reading habits has more chances of success. In other words, sound knowledge on different topics like politics, finance, economy, science and technology is helpful.

◎ Power to convince effectively is another quality that makes you stand out among others.

◎ Clarity in speech and expression is yet another essential quality.

◎ If you are not sure about the topic of discussion, it is better not to initiate. Lack of knowledge or wrong approach creates a bad impression. Instead, you might adopt the wait-and-watch attitude. Listen attentively to others, may be you would be able to come up with a point or two later.

◎ A GD is a formal occasion where slang is to be avoided.

◎ A GD is not a debating stage. Participants should confine themselves to expressing their viewpoints. In the second part of the discussion candidates can exercise their choice in agreeing, disagreeing or remaining neutral.

◎ Language use should be simple, direct and straightforward.

◎ Don't interrupt a speaker when the session is on. Try to score by increasing your size, not by cutting others short.

◎ Maintain rapport with fellow participants. Eye contact plays a major role. Non-verbal gestures, such as listening intently or nodding while appreciating someone's viewpoint, speak of you positively.

◎ Communicate with each and every candidate present. While speaking don't keep looking at a single member. Address the entire group in such a way that everyone feels you are speaking to him or her.

How to Arrange for a Discussion?

1. The subject is selected.

2. Prepare a list of all who are going to participate—only competent persons.

3. Specific areas for discussion are identified and entrusted to a person such as Etiology to one person, Epidemiology to another person and so on.

4. Arrange the discussion in groups so that each person can see the other person. The arrangement should be square or circular so that all can see each other.

5. There should be a leader or a Chairperson. There should be sufficient space between the chairperson and others.

6. All persons should be seated, and there should be the identity of the person in front (name board). Discussion can be started after making everybody comfortable.

7. Introduce each other and it is leader's duty to learn the names of members. Call only by name.

8. Start on time and close at prearranged time.

9. It is the duty of the leader to involve all the persons participating in the discussion. Even if they feel shy, make them talk.

10. Limit the individual contribution to the barest minimum time and as a leader, be conversant yourself or familiarize yourself with all aspects of the problems.

11. To stimulate discussion, put forward leading questions (in general and not to specific individuals; if so, it becomes like any oral communication).

12. As a leader, you are free to interrupt the speech makers in the discussion group.

13. Go from the discussion to investigations. Ideas obtained in the discussion should be put into practice. Investigation is the only indication that your thinking had been stimulated.

14. If there is no such follow-up, attending the discussion had been a waste.

15. There may be some unanswered questions—some areas may be left without any adequate information. These should be highlighted at the end and left for discussion later.

16. The chairperson's duty is to summarize the proceedings of the discussion at the end and spell out the conclusions.

GENERAL MEETINGS, WORKSHOPS AND SEMINARS

How to Conduct a Scientific Session

Scientific sessions are held generally for discussing scientific matters. The Chairpersons should be competent specialists in their field. They are responsible for conducting and concluding the seminar. His responsibilities are different from others participating in the session. They should be conscious of the time, and the credit goes to them for the effectiveness of communication.

1. Sessions should be started in time. Need not wait for others to come.

2. All speakers should be given time to speak.

3. The topic should be introduced before start of session.

4. Ideas and implications should be stated positively.

5. Impress upon the audience the importance of the meeting and how it is worthwhile to have discussion on the topic in the present situation. Need for specific conclusions should be stressed.

6. The leadership is an active force and should exert itself to guide, direct, restrict, develop, contain, expand and stimulate thinking of the research group. Remain impartial as well.

7. Watch the passing of the meeting. Keep ideas moving.

8. Leader can express non-committal ideas when the discussion attains a standstill.

9. Watch for emotional build-ups. Emotions do not solve the problems. Make rational decisions. Use humour to lighten the situation. Point out that both the (fighting) members may be right (without involving yourself in trouble).

10. Bringing in background information relevant to the topic is the responsibility of the leader. Help in drawing logical conclusions.

11. See that only one person talks at a time.

12. "Private discussions" within a group should be discouraged.

13. Draw contributions by praising the group as a unit. Successful and workable decisions are the signs of successful meeting.

14. Guide the meetings from problems to solutions.

The Leader

Most of the meetings end as a failure due to lack of direction. This places the responsibility on the leader. A wrong emphasis, a misinterpretation of a contribution, a missed fact, all can lead to discussion off the track. Clarify the contribution. Make sure that what was said by one member is understood by all present.

1. Ask questions in areas that may have been missed.

2. Pull out the details to make each suggestion crystal clear.

3. Define the words that will lead to misinterpretation.

4. Members may be asked to elaborate the points if required.

5. Make frequent summaries during the meeting. Summaries, are in fact, a running score card of the discussions that are going on. Summaries will help to check on disagreements and point out to the members what conflicts demand. Quick summaries will not interrupt the flow of the meeting.

6. Introduce humour to reduce the tension.

7. At the end of the meeting, the leader should be satisfied that the discussion brought out specific conclusions. The major disagreements and the future plans to be undertaken should be discussed in the meeting. A clear statement of the decisions is

an important factor. Point out the disagreements very clearly and evaluate them as to how much they are significant.

Generally the leader is helped by the rapporteurs in the conduct of seminar.

How to Select a Topic for Presentation?

1. Choose a subject that suits speaker, audience and the occasion.
2. Choose a subject that suits the literature available.
3. Bear in mind that it is the speaker who is teaching the audience.
4. The topic should be interesting and should convey specific information.
5. It should be simple, and clarity is important.
6. The audience should leave with some knowledge.
7. It should motivate the thinking of the audience.
8. The topic should be convincing.
9. The presentation should be supported by statistics.
10. The audience must expand and exploit your ideas further.

GLOSSARY

Abstract A short summary or version prepared by cutting short a larger work. A statement summarizing the important points of a text.

Acknowledgement An expression of thanks or a token of appreciation.

Aperture An opening, such as a hole, gap, or slit. Usually an adjustable opening in an optical instrument, such as a camera or telescope, that limits the amount of light passing through a lens or on to a mirror.

Apostrophe The superscript sign (') used to indicate the omission of a letter or letters from a word, the possessive case.

Appendix A collection of supplementary material, usually at the end of a book.

Attrition A gradual diminution in number or strength because of constant stress.

Basic/Fundamental research Research carried out to increase understanding of fundamental principles. Sometimes called pure research.

Bias A statistical sampling or testing error caused by systematically favouring some outcomes over others.

Bibliography A list of the works of a specific author or publisher. The description and identification of the editions, dates of issue, authorship, and typography of books or other written material.

Cartogram A presentation of statistical data in geographical distribution on a map.

Case-control study A widely used method of observational epidemiological study, and is an application of medical history-taking that aims to identify the cause of disease among a group of people, or the cause–effect relationships of a condition of interest.

Catalogue A list or itemized display, as of titles, course offerings, or articles for exhibition or sale, usually including descriptive information or illustrations.

Cluster sampling A random sampling plan in which the population is subdivided into groups called clusters so that there is small variability within clusters and large variability between clusters.

Cohort study A longitudinal study of the same group of people (the cohort) over time.

Colon A punctuation mark (:) used after a word introducing a quotation, an explanation, an example, or a series.

Comma A punctuation mark (,) used to indicate a separation of ideas or of elements within the structure of a sentence.

Conceptual research Research relating to concepts or mental conception.

Confidence interval A statistical range with a specified probability that a given parameter lies within the range.

Convenience sampling Samples of a population that are chosen based on their relative ease of access.

Cross-sectional study The scientific method for the analysis of data gathered from two or more samples at one point in time.

Database A collection of data arranged for ease and speed of search and retrieval. Also called data bank.

Descriptive research Research that describes data and characteristics about the population or phenomenon being studied. Descriptive research answers the questions *who*, *what*, *where*, *when* and *how*.

Dissertation A lengthy, formal treatise, especially one written by a candidate for the doctoral degree at a university; a thesis.

Empirical research Any research that bases its findings on direct or indirect observation as its test of reality.

Encyclopaedia A comprehensive reference work containing articles on a wide range of subjects or on numerous aspects of a particular field, usually arranged alphabetically.

Error The difference between a computed or measured value and a true or theoretically correct value.

Exclamation mark A punctuation mark (!) used after an exclamation.

Exploratory research A type of research conducted because a problem has not been clearly defined. Exploratory research helps determine the best research design, data collection method and selection of subjects.

Footnote A note placed at the bottom of a page of a book or manuscript and comments on or cites a reference for a designated part of the text.

Frequency curve A graphical representation of a continuous frequency distribution; the value of the variable is the abscissa and the frequency is the ordinate.

Frequency polygon A graph obtained from a frequency distribution by joining with straight lines points whose abscissae are the midpoints of successive class intervals and whose ordinates are the corresponding class frequencies.

Full stop or period A punctuation indicating the end of a sentence.

Google A trademark used for an Internet search engine.

Graph A diagram that exhibits a relationship, often functional, between two sets of numbers as a set of points having coordinates determined by the relationship. Also called plot.

Histogram A bar graph of a frequency distribution in which the widths of the bars are proportional to the classes into which the variable has been divided and the heights of the bars are proportional to the class frequencies.

Hyphen A punctuation mark (-) used between the parts of a compound word or name or between the syllables of a word, especially when divided at the end of a line of text.

Hypothesis A tentative explanation for an observation, phenomenon, or scientific problem that can be tested by further investigation.

Index An alphabetized list of names, places, and subjects treated in a printed work, giving the page or pages on which each item is mentioned.

Interrogation/question mark A punctuation symbol (?) written at the end of a sentence or phrase to indicate a direct question. Also called interrogation point.

Longitudinal research The study of a group of individuals at regular intervals over a relatively long period of time.

Longitudinal study Epidemiological studies that record data from a representative sample at repeated intervals over an extended span of time rather than at a single or limited number over a short period.

Mean The average value of a set of numbers.

Method A means or manner of procedure, especially a regular and systematic way of accomplishing something.

Methodology The theoretical analysis of the methods appropriate to a field of study or to the body of methods and principles particular to a branch of knowledge.

Multistage sampling A sampling method in which the population is divided into a number of groups or primary stages from which samples are drawn; these are then divided into groups or secondary stages from which samples are drawn, and so on.

Offline Not connected to a computer or computer network.

Ogives A frequency distribution curve in which the frequencies are cumulative.

Online Connected to a computer or computer network. Accessible via a computer or computer network.

Panel study A longitudinal study of the same group of people over time. By contrast with a cohort study, the panel study gathers information on the same people at each time point.

Parenthesis Either or both of the upright curved lines, ( ), used to mark off explanatory or qualifying remarks in writing or printing or enclose a sum, product, or other expression considered or treated as a collective entity in a mathematical operation.

Periodical A publication issued at regular intervals of more than one day.

Pictograph A picture representing a word or idea; A pictorial representation of numerical data or relationships, especially a graph, but having each value represented by a proportional number of pictures.

Pie graph A circular diagram, sectioned by radial lines, used to represent distributions within a whole.

Preface A preliminary statement or essay introducing a book that explains its scope, intention, or background and is usually written by the author.

Preliminary Prior to or preparing for the main matter, action, or business; introductory or prefatory. Something that precedes, prepares for, or introduces the main matter.

Probability sampling A method of sampling from a finite population where the probability of each set of units being selected is known.

Proofread To read (a copy or proof) in order to find errors and mark corrections.

Prospective Likely or expected to happen. Anticipated.

Purposive sampling A form of sampling in which the selection of the sample is based on the judgment of the researcher as to which subjects best fit the criteria of the study.

Questionnaire A form containing a set of questions, especially one addressed to a statistically significant number of subjects as a way of gathering information for a survey.

Quota sampling A method of sampling in a survey in which the interviewer is instructed to include certain prescribed percentages of people who come from various identifiable subpopulations, e.g. men over 60, women under 30, unemployed men.

Quote To repeat or copy the words of another person/source, usually with acknowledgment of the source.

Random sampling A sampling from some population where each entry has an equal chance of being drawn.

Randomization The process of randomly allocating the experimental units across the treatment groups. Thus, if the experiment compares a new drug against a standard drug used as a control, the patients should be allocated to new drug or control by a random process.

Reliability The amount of credence placed in a result. The precision of a measurement, as measured by the variance of repeated measurements of the same object.

Replication The repetition of an experimental condition so that the variability associated with the phenomenon can be estimated.

Research The study directed towards the increase of knowledge, the primary aim being a greater knowledge or understanding of the subject under study.

Retrospective Looking back on, contemplating, or directed to the past.

Sample A portion, piece, or segment that is representative of a whole. A set of elements drawn from and analysed to estimate the characteristics of a population.

Sampling frame The full list of members of the population to be studied from which a sample can be drawn.

Search engine A software program that searches a database and gathers and reports information that contains or is related to specified terms.

Semicolon A mark of punctuation (;) used to connect independent clauses, and indicating a closer relationship between the clauses than a period does.

Stratified random sampling A random sample of specified size that is drawn from each stratum of a population.

Summary   A shortened version of the original. The main purpose of such a simplification is to highlight the major points from the genuine (much longer) subject, e.g. a text, a film or an event. The target is to help the audience get the gist in a short period of time.

Survey   A detailed inspection or investigation. A gathering of a sample of data or opinions considered to be representative of a whole.

Systematic sampling   The selection of every k^{th} element from a sampling frame, where k, the sampling interval, is calculated using the formula, $k = \dfrac{\text{Population size } (N)}{\text{Sample size } (n)}$. Using this procedure, each element in the population has a known and equal probability of selection.

Thesis   A dissertation advancing an original point of view as a result of research, especially as a requirement for an academic degree.

Trend study   The most common longitudinal study among others. A trend study samples *different* groups of people at different points in time from the *same* population.

Validity   Correctness; especially the degree of closeness by which iterated results approach the correct result.

Web crawler   A program or automated script that browses the world wide web in a methodical, automated manner.

Yahoo   A trademark used for an Internet search engine.